Digital Control
in Power Electronics

2nd Edition

Synthesis Lectures on Power Electronics

Editor
Jerry Hudgins, *University of Nebraska–Lincoln*

Synthesis Lectures on Power Electronics will publish 50- to 100-page publications on topics related to power electronics, ancillary components, packaging and integration, electric machines and their drive systems, as well as related subjects such as EMI and power quality. Each lecture develops a particular topic with the requisite introductory material and progresses to more advanced subject matter such that a comprehensive body of knowledge is encompassed. Simulation and modeling techniques and examples are included where applicable. The authors selected to write the lectures are leading experts on each subject who have extensive backgrounds in the theory, design, and implementation of power electronics, and electric machines and drives.

The series is designed to meet the demands of modern engineers, technologists, and engineering managers who face the increased electrification and proliferation of power processing systems into all aspects of electrical engineering applications and must learn to design, incorporate, or maintain these systems.

Digital Control in Power Electronics, 2nd Edition
Simone Buso and Paolo Mattavelli
2015

Transient Electro-Thermal Modeling of Bipolar Power Semiconductor Devices
Tanya Kirilova Gachovska, Bin Du, Jerry L. Hudgins, and Enrico Santi
2013

Modeling Bipolar Power Semiconductor Devices
Tanya K. Gachovska, Jerry L. Hudgins, Enrico Santi, Angus Bryant, and Patrick R. Palmer
2013

Signal Processing for Solar Array Monitoring, Fault Detection, and Optimization
Mahesh Banavar, Henry Braun, Santoshi Tejasri Buddha, Venkatachalam Krishnan, Andreas Spanias, Shinichi Takada, Toru Takehara, Cihan Tepedelenlioglu, and Ted Yeider
2012

The Smart Grid: Adapting the Power System to New Challenges
Math H.J. Bollen
2011

Digital Control in Power Electronics
Simone Buso and Paolo Mattavelli
2006

Power Electronics for Modern Wind Turbines
Frede Blaabjerg and Zhe Chen
2006

Digital Control in Power Electronics, 2nd Edition

Simone Buso and Paolo Mattavelli

ISBN: 978-3-031-01371-3   paperback
ISBN: 978-3-031-02499-3   ebook

DOI 10.1007/978-3-031-02499-3

A Publication in the Springer series
SYNTHESIS LECTURES ON POWER ELECTRONICS

Lecture #7
Series Editor: Jerry Hudgins, *University of Nebraska–Lincoln*
Series ISSN
Print 1931-9525   Electronic 1931-9533

Digital Control
in Power Electronics

2nd Edition

Simone Buso
University of Padova

Paolo Mattavelli
University of Padova

SYNTHESIS LECTURES ON POWER ELECTRONICS #7

ABSTRACT

This book presents the reader, whether an electrical engineering student in power electronics or a design engineer, a selection of power converter control problems and their basic digital solutions, based on the most widespread digital control techniques. The presentation is primarily focused on different applications of the same power converter topology, the half-bridge voltage source inverter, considered both in its single- and three-phase implementation. This is chosen as the test case because, besides being simple and well known, it allows the discussion of a significant spectrum of the most frequently encountered digital control applications in power electronics, from digital pulse width modulation (DPWM) and space vector modulation (SVM), to inverter output current and voltage control, ending with the relatively more complex VSI applications related to the so called *smart-grid* scenario. This book aims to serve two purposes: (1) to give a basic, introductory knowledge of the digital control techniques applied to power converters; and (2) to raise the interest for discrete time control theory, stimulating new developments in its application to switching power converters.

KEYWORDS

digital control in power electronics, discrete time control theory, half-bridge voltage source converters, power converters, power electronics

Digital Control in Power Electronics is dedicated to the memory of Professor Luigi Malesani, the founder of the PEL group and our unforgettable maestro.

Contents

Preface for the Second Edition

After more than eight years from its first publication, we felt it was about time to revise and update the material presented in this book. In preparing the revision, we did not want to change the introductory, concise style of the first edition, that, in our opinion, has been, and still is, one of the strong sides of the book. Yet, we wanted to provide the reader with an updated, fresher image of digital control applications in power electronics.

A few of the trends that had been identified some years ago, such as the evolution of digital controllers from software to hardware implementations, have indeed become standard practice, while new, partially unforeseen, application fields have emerged, like, among all, the distributed generation and "smart-grid" ones. Without trying to cover these topics exhaustively, which would have gone beyond the introductory scope of the book, we still took the chance to present, with this second edition, new information and experimental data from our more recent studies related to them.

In doing so, we have compiled a couple of new chapters, one dedicated to the multi-sampled versions of the basic current controller implementations, that represented the core of the first edition, and the second to the complex digital control architectures that characterize the above mentioned, emerging application fields. In addition, we have decided to revise the original discussion of basic current controllers, essentially to provide experimental verification data, corroborating the original simulation-based exemplifications, and, in our opinion, improving the quality of the presentation.

Simone Buso and Paolo Mattavelli
April 2015

Acknowledgments

This book has been written on the basis of the authors' expertise, that has been almost fully acquired within the power electronics research group (PEL group) of the University of Padova. Without the help and dedication of the people that have worked and are still working in the team, the writing of the book would have been altogether impossible. Therefore, the authors would like to acknowledge and thank everybody at PEL group for the time spent together in creating, discussing, and testing all the ideas presented in this book.

Simone Buso and Paolo Mattavelli
April 2015

CHAPTER 1

Introduction

Power electronics and discrete time system theory have been closely related to each other from the very beginning. This statement may seem surprising at first, but, if one thinks of switch mode power supplies as *variable structure periodic* systems, whose state is determined by *logic signals*, the connection becomes immediately clearer. A proof of this may also be found in the first, fundamental technical papers dealing with the analysis and modeling of pulse width modulated power supplies or peak current mode controlled DC-DC converters: they often provide a mathematical representation, of both the switching converters and the related control circuits, resembling or identical to that of *sampled data* dynamic systems.

This fundamental contiguousness of the two apparently far areas of engineering is probably the strongest, most basic motivation for the considerable amount of research that, over the years, has been dedicated to the application of digital control to power electronic circuits. From the original, basic idea of implementing current or voltage controllers for switching converters using digital signal processors or microcontrollers, that represents the foundation of all current industrial applications, the research focus has moved to more sophisticated approaches, where the design of custom integrated digital controllers is no longer presented just as an academic curiosity, but is rather perceived like a sound, viable solution for the next generation of high performance power supplies.

Besides that, new technologies, like the hardware programmable logic circuits and the fast analog to digital converters currently available, are spurring a thorough revision of the standard industrial practice, as more and more advanced, multi platform controller organizations have become widely affordable.

Referring to this fairly complex scenario, the purpose of this book is actually twofold. In the first place, we would like to introduce the reader to basic control problems in power electronic circuits and to illustrate their more classical, widely applied digital solutions. In addition to that, on the occasion of its second edition, we would like to offer the reader a glimpse of the most recent advancements in digital control applications, presenting both a selection of possible improvements of the basic control strategies and the new implementation techniques that are now making their appearance both in academic research and in the industrial practice. We hope this will serve two purposes: on the one hand, to give a basic, introductory knowledge of the digital control techniques applied to power converters; on the other hand, to raise the interest for discrete time control theory, hopefully stimulating new, further developments in its application to power converters and power systems.

MODERN POWER ELECTRONICS

Classical power electronics may be considered, under several points of view, a mature discipline. The technology and engineering of discrete component-based switch mode power supplies are nowadays fully developed industry application areas, where one does not expect to see any outstanding innovation, at least in the near future. Symmetrically, at the present time, the research fields concerning power converter topologies and the related conventional, analog control strategies seem to have been almost thoroughly explored. A notable exception might be represented by multi-level, multi-phase inverter topologies, that are still being intensively investigated. They could indeed play a key role in medium to high voltage power electronics applications, required, as an example, by the expected renovation of the electrical power distribution grids, according to the so called "smart-grid" concept.

On the other hand, we can identify a few very promising fields where power electronics research is still moving on the cutting edge of innovation. For example, large bandgap semiconductor devices, in particular the semiconductor technologies based on silicon carbide (SiC), gallium arsenide (GaAs), and gallium nitride (GaN) are nowadays proving to be practically usable not just for ultra high frequency amplification of radio signals or as the power LED technology basic materials, but also for switching power conversion applications. This is really enabling the industrial production of high frequency (multi-MHz) and/or high temperature power converter circuits and, consequently, a very significant leap in the achievable power densities. A significant impact of these new technologies is expected to be seen in the transportation field, both for automotive applications (with particular reference to the increasing diffusion of electric vehicles) and for the aerospace ones (within the "more electric aircraft" framework).

But the rush for higher and higher power densities motivates research also in other directions. Among these, we would like to mention a few that, in our vision, will continue to play a very significant role for quite a long time.

The first is the integration in a single device of magnetic and capacitive passive components, that may allow the implementation of minimum volume, quasi monolithic, converters. Equally important is the analysis and mitigation of electromagnetic interference (EMI), that is becoming more and more fundamental for the design of compact, high-frequency, converters, where critical auto-susceptibility problems are often encountered. The third one is the development of technologies and design tools allowing the integration of control circuits and power devices on the same semiconductor chip, according to the so-called *smart power* concept. Finally, reliability studies are going to assume increasing importance, as a result of the ongoing widespread penetration of power electronics in modern society.

These research areas represent good examples of what, in our vision, can be considered modern power electronics. From this standpoint, the application of digital control techniques to switch mode power supplies can play a very significant role. Indeed, the integration of complex control functions, as those that are likely to be required by the next generation power supplies, is a problem that can realistically be tackled only with the powerful tools of digital control design.

WHY DIGITAL CONTROL

The application of digital control techniques to switch mode power supplies has always been considered very interesting, mainly because of the several advantages a digital controller shows, when compared to an analog one.

Surely, one of the most relevant is the possibility it offers of implementing sophisticated control laws, taking care of nonlinearities, parameter variations or construction tolerances by means of self-analysis, and auto-tuning strategies, very difficult or impossible to implement analogically.

Another very important advantage is the flexibility inherent in any digital controller, that allows the designer to modify the control strategy, or even to totally re-program it, without the need for significant hardware modifications. Also very important are the higher tolerance to signal noise and the complete absence of aging effects or thermal drifts.

In addition to all this, we must consider that, nowadays, a large variety of electronic devices, from home appliances to industrial instrumentation, require the presence of some form of human-machine interface (HMI). Its implementation is almost impossible without having some kind of embedded microprocessor. The utilization of the computational power, that thus becomes available, also for lower level control tasks is almost unavoidable.

For these reasons, the application of digital controllers has been increasingly spreading and has become the only effective solution for a whole lot of industrial power supply production areas. To give an example, adjustable speed drives (ASDs) and uninterruptible power supplies (UPSs) are nowadays fully controlled by digital means.

The increasing availability of low-cost, high-performance, microcontrollers and digital signal processors stimulates the diffusion of digital controllers also in areas where the cost of the control circuitry is a truly critical issue, like that of power supplies for portable equipment, battery chargers, electronic welders, and several others. However, a significant increase of digital control applications in these very competing markets is not likely to take place until new implementation methods, different from the traditional microcontroller or DSP unit application, prove their viability.

From this standpoint, the research efforts towards digital control applications need to be focused on the design of custom integrated circuits, more than on algorithm design and implementation. Issues like occupied area minimization, scalability, power consumption minimization, and limit cycle containment play a key role. The power electronics engineer is, in this case, deeply involved in the solution of digital integrated circuit design problems, a role that will be more and more common in the future.

As an intermediate step, the application of hardware programmable devices, such as field programmable gate array (FPGA) circuits, is already showing several advantages with respect to the standard implementation methods. In a sense, it allows to merge the flexibility of software programmable devices with the high performance of custom digital controllers. Besides, it allows the designer to avoid, at least to some extent, the complications of a digital chip design, guaranteeing limited development times for new applications. This is the domain where the distance between algorithms and digital circuits is really minimal, a condition that offers a whole lot of new opportunities for digital control applications.

TRENDS AND PERSPECTIVES

From the above discussion, it will be no surprise if we say that we consider the increasing diffusion of digital control in power electronics virtually unstoppable. The advantages of the digital control circuits, as we have briefly outlined in the previous section, are so evident that, in the end, all the currently available analog integrated control solutions are going to be replaced by new ones, embedding some form of digital signal processing core. Indeed, it is immediate to recognize that the digital control features perfectly match the needs of present and, even more, future, highly integrated, power converters. The point is only how long this process is going to take. We can try to outline the future development of digital controllers distinguishing the different application areas.

The medium to high power applications, like electrical drives, test power supplies, uninterruptible power supplies, renewable energy source interfaces, and grid tied inverters for smart grids are likely to be developed according to the same basic hardware organization for a long time to come. The application of microcontroller units or digital signal processors in this area is likely to remain very intensive, although their coupling to hardware programmable logic devices is already taking place and will be even stronger in the future. A further evolution trend is indeed represented by the increasing integration of higher level functions, e.g., those concerning communication protocols for local area networks or field buses, complex human-machine interfaces, remote diagnostic capabilities, that currently require the adoption of different signal processing units, with low-level control functions.

As far as the low-power applications are concerned, as we mentioned in the previous section, the market for integrated digital controllers has not yet fully developed. However, the application of digital control in this field is the object of an intensive research. In the near future, new control solutions can be anticipated, that will replace analog controllers with equivalent digital systems, in a way that can be considered almost transparent to the user. Successively, the complete integration of power and control circuitry is likely to determine a radical change in the way low power converters are designed.

WHAT IS IN THIS BOOK

As mentioned above, in front of the complex and exciting perspectives for the application of digital control to power converters, we decided to aim this book primarily at giving the reader a basic and introductory knowledge of some typical power converter control problems and of their digital solutions. Referring to the above discussion, we decided to dedicate the largest part of our presentation to topics that can be considered the current state of the art for industrial applications of digitally controlled power supplies.

This book is consequently proposed to power electronics students, or designers, who would like to have an overview of the most widespread digital control techniques. It is not intended to provide an exhaustive description of all the possible solutions for any considered problem, nor to describe the more recent research advances related to any of them. This choice has allowed us to keep the presentation of the selected materials relatively agile and to give it an immediate, practical usefulness.

Accordingly, what the reader should know to take full advantage of the contents that are here presented is relatively little: a basic knowledge of some power electronic circuits (essentially half-bridge and full-bridge voltage source inverters) and of the fundamental mathematical tools that are commonly employed in modeling continuous and discrete time dynamic systems (Laplace transform and $\mathcal{Z}$ transform, for starters) will perfectly do.

As the reader will realize, if he or she will have the patience to follow us, the book is conceived to explain the different concepts essentially by means of examples. To limit the risk of being confusing, proposing several different topologies, we decided to take into account a single, relatively simple case and to develop its analysis all along the text. Doing so, the contents we have included allowed us to present, organically and without too many context changes, a significant amount of control techniques and related implementation details.

In summary, the book is organized as follows. Chapter 2 describes the considered test case, a voltage source inverter, and the first control problem, i.e., the implementation of a current control loop, discussing in the first place its analog, i.e., continuous time, solutions.

Chapter 3 is dedicated to digital control solutions for the same problem: in the beginning we present a relatively simple one, i.e., the discretization of continuous time controllers. In the following, other fully digital solutions, like those based on discrete time state feedback and pole placement, are presented.

In Chapter 4 the implementation of multi-sampled versions of the current controllers presented in Chapter 3 is discussed. The achievable performance is evaluated both in comparison with the conventional controllers' one and in absolute terms, giving the reader a complete view of both small- and large-signal characteristics.

Chapter 5 is dedicated to the extension to three phase systems of the solutions presented for the single-phase inverter. In this chapter we discuss Space Vector Modulation (SVM) and rotating reference frame current controllers, like those based on Park's transformation.

Chapter 6 presents the implementation of external control loops, wrapped around the current controller, in what is typically known as a multi-loop controller organization. The design of an output voltage controller, as is needed in uninterruptible power supplies, is first considered. Both large bandwidth control strategies and narrow bandwidth ones, based on the repetitive control concept, are analyzed. After that, two other significant examples of multi-loop converter control, that we may find in controlled rectifiers and active power filters, are considered and briefly discussed.

Finally, in Chapter 7, the control of distributed energy resources and storage devices, typical of the so-called smart grids, is taken as an example to illustrate the ultimate evolution of VSI digital controllers. These are here required to simultaneously perform a large variety of functions, from low level ones, i.e., modulation and current control, to middle level ones, such as output voltage or injected power control, and high level ones, like grid optimization, communication protocol support, or data logging. The organization of such different functions into a multiple layer stack of hardware and software programmable devices is illustrated in this chapter, where the fundamental tools and methodologies that enable fast and reliable design procedures are also reviewed.

CHAPTER 2

The Test Bench: A Single-Phase Voltage Source Inverter

The aim of this chapter is to introduce the test bench converter we will be dealing with in the whole following presentation. As mentioned in the introduction, it would be extremely difficult to describe the numerous applications of digital control to switch mode power supplies, since this is currently employed in very wide variety of cases. In order not to confuse the reader with a puzzle of several different circuit topologies and related controllers, what we intend to do is to consider just a single, simple application example, where the basics of the more commonly employed digital control strategies can be effectively explained. Of course, the concepts we are going to illustrate, referring to our test case, can find a successful application also to other converter topologies.

The content of this chapter is made up, in the first place, by an introductory, but fairly complete, description of the power converter we will be discussing all over this book, i.e., the half bridge voltage source inverter. Secondly, the principles of its more commonly adopted low level control strategy, namely Pulse Width Modulation (PWM), will be explained, at first in the continuous time domain, successively in the discrete time domain. The issues related with PWM control modeling are fundamental for the correct formulation of a Switch Mode Power Supply (SMPS) digital, or even analog, control problem, so that this part of the chapter can be considered essential to the understanding of everything that follows. The final part of the chapter is instead dedicated to a summary of the more conventional analog control strategies, that will serve as a reference for all the following developments.

2.1 THE VOLTAGE SOURCE INVERTER

The considered test bench converter circuit is shown in Fig. 2.1. As can be seen, the power converter we want to take into consideration is a single-phase Voltage Source Inverter (VSI). The VSI has a conventional topological structure, that is known as a *half bridge*. We will analyze the power converter's organization in some detail now.

2.1.1 FUNDAMENTAL COMPONENTS

The ideal voltage sources V_{DC} at the input are, in practice, approximately implemented by means of suitably sized capacitors, fed by a primary energy source. They are normally large enough to store a considerable amount of energy and their purpose is to deliver it to the load, rapidly enough

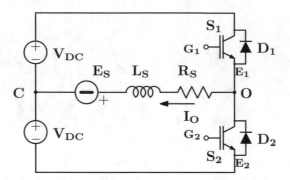

Figure 2.1: Simplified schematic of a half-bridge voltage source inverter.

not to cause the circulation of substantial high-frequency currents through the primary source. This, in turn, can be represented by any real DC voltage source, from batteries to line-fed rectifiers, depending on the particular application. However, for our discussion, modeling the energy source as an ideal voltage source does not represent any limitation.

The power switches are represented with the conventional IGBT symbol, but it is possible to find implementations with very different switch technologies, such as, power MOSFETs or, for very high power application, thyristors. As can be seen, each switch is paralleled to a *free-wheeling* diode, whose purpose is to make the switch bi-directional, at least as far as the current flow is concerned. This interesting property makes the VSI of Fig. 2.1 a four-quadrant converter, with the capability of both delivering and absorbing power.

Again, in order to simplify the treatment of our control problems and without any loss of generality, we will assume that the switch plus diode couple behaves like an ideal switch, i.e., one whose voltage is zero in the "on" state and whose current is zero in the "off" state. Moreover, we will assume the transition from the "on" state to the "off" state and vice versa takes place in a null amount of time. A more detailed discussion of real life switches' behavior will be provided further on in this chapter.

In our simple example, the load will be described as the series connection of a resistor R_S, an inductor L_S, and a voltage source E_S, that can be either DC or AC. We will learn to control the current across the load using several different strategies. It is worth mentioning that, with this particular structure, the load model is capable of representing various different applications of the VSI, including electrical drives, voltage-controlled current sources, controlled rectifiers. The role and meaning of the different components, in particular of voltage E_S, will be different in each case, but the structure will be exactly the same.

2.1.2   REQUIRED ADDITIONAL ELECTRONICS: DRIVING AND SENSING

Several components are needed to allow the proper operation of the VSI that were not described in the previous section. First of all, the power switches need to be driven by a suitable control circuit, allowing the controlled commutation of the device from the "on" to the "off" state and vice versa. Depending on the particular switch technology, the driving circuitry will have different implementations. For example, in the case of MOSFET or IGBT switches the driving action consists in the charging and discharging of the device input capacitance, which is, in fact, a power consuming operation. To take care of that, suitable drivers must be adopted, whose input is represented by the logic signals determining the desired state of the switch and output is the power signal required to bring the switch into that state. A typical complication in the operation of drivers is represented by the floating control terminals of the high side switch (G_1 and E_1 in Fig. 2.1). Controlling the voltage between those terminals and, simultaneously, that between the same terminals of the low side switch (G_2 and E_2 in Fig. 2.1) requires the adoption of isolated driving circuits or the generation of floating power supplies, e.g., based on bootstrap capacitors.

We will not discuss further the operation of these circuits and simply assume that the logic state of the control signal is instantaneously turned into a proper switch state. An exception to this will be the discussion of dead-times, presented in the following. Of course, the interested reader can find more details regarding state-of-the art switch drivers in technical manuals or datasheets, easily available on the World Wide Web, like, for example, [1].

Besides drivers, the controlled operation of the converter requires the measurements of several electrical variables. Typically, the input voltage of the inverter circuit, V_{DC}, its output current, i.e., the current flowing through the load, I_O, and, sometimes, the voltage E_S are measured and used in the control circuit. The acquisition of those signals requires suitable signal conditioning circuits, analog in nature, that can range from simple resistive voltage dividers and/or current shunts, possibly combined to passive filters, to more sophisticated solutions, for example those employing operational amplifiers, to implement active filters and signal scaling, or Hall sensors, to measure currents without interfering with the power circuit.

In our discussion we will simply assume that the required control signals are processed by suitable conditioning circuits that, in general, will apply some scaling and filtering to each electrical variable. The frequency response of the acquisition filters and the scaling factors implied by sensors and conditioning circuits will be properly taken into account in the controller design examples we will present in the following chapters.

2.1.3   PRINCIPLE OF OPERATION

The principle of operation of the half bridge inverter of Fig. 2.1 is the following. Closing the high-side switch S_1 imposes a voltage across the load (i.e., V_{OC} in the figure) equal to $+V_{DC}$. In contrast, closing the low side switch S_2 imposes a voltage $-V_{DC}$ across the load. If a suitable control circuit regulates the *average* voltage across the load (see Section 2.1.4 for a rigorous definition of average load voltage) between these two extremes, it is clearly possible to make the state variable I_O follow any desired trajectory, provided that this is consistent with the physical limitations imposed by the topology. The main limitation is obvious: the voltage across the load cannot exceed $\pm V_{DC}$. Other limitations can be seen, giving just a little closer look to the circuit. Considering, as an example, the particular case where E_S and R_S are both equal to zero, the I_O current will be limited in its variations, according to the following equation:

$$\left| \frac{d I_O}{dt} \right| \leq \frac{V_{DC}}{L_S}. \tag{2.1}$$

In practice, the maximum current absolute value will be limited as well, mainly because of the limited current handling capability of the active devices. This limitation, differently from the previous ones, is not inherent to the circuit topology and will need to be enforced by a current controller, in order to prevent accidental damage to the switches, for example in the case of a short circuit in the load. What should be clear by now is that any controller trying to impose voltages, currents, or current rates of change beyond the above-described limits will not be successful: the limit violation will simply result in what is called *inverter saturation*. It is worth adding that, in our following discussion, we will consider linear models of the VSI, capable of describing its dynamic behavior in a *small-signal* approximation. Events like inverter saturation, typical of large-signal inverter operation, will not be correctly modeled. In order to further clarify these concepts, the derivation of a small-signal linear model for the VSI inverter of Fig. 2.1 is presented in the Aside 1.

In the most general case, the VSI controller is organized hierarchically, in a multi-level, or multi-layer, architecture. At the bottom layer a controller determines the state of each of the two switches, and in doing so, the average load voltage. This layer is called the *modulator* layer. The strategy according to which the state of the switches is changed along time is called the *modulation law*. The input to the modulator is the set-point for the load average voltage, normally provided by a higher-level control loop. A direct control of the average load voltage is also possible: in this case the VSI is said to operate in open loop conditions. However, this is not a commonly adopted mode of operation, since no control of load current is provided in the presence of load parameter variations.

Because of that, in the large majority of cases, a current controller can be found immediately above the modulator layer. This is responsible for providing the set-point to the modulator. Similarly, the current controller set-point can be provided by a further external control loop or directly by the user. In the latter case, the VSI is said to operate in *current mode*, meaning that the control circuit has turned a voltage source topology into a controlled current source. We will deal with further external control loops in one of the following chapters; for now, we will focus on the modulator and current control layers.

Indeed, the main purpose of this chapter is exactly to explain how these two basic controller layers are organized and how the current regulators can be properly designed.

2.1.4 DEAD-TIMES

Before we move to describe the modulator layer, one final remark is needed to complete the explanation of the VSI operation. The issue we want to address here is known as *switching dead-time*. It is evident from Fig. 2.1 that in no condition the simultaneous conduction of both switches should be allowed. This would indeed result into a short circuit across the input voltage sources, leading to an uncontrolled current circulation through the switches and, very likely, to inverter fatal damage. Any modulator, whatever its implementation and modulation law, should be protected against this event. In the ideal switch hypothesis of Section 2.1.1, the occurrence of switch cross conduction can be easily prevented by imposing, under any circumstances, logically complementary gate signals to the two switches. Unfortunately, in real life cases, this is not a sufficient condition to avoid cross conduction. It should be known from basic power electronics knowledge that real switch commutations require a finite amount of time and that the commutation time is a rather complex function of several variables among which are the commutated current and voltage, the gate drive current, the device junction temperature and so on. It is therefore impossible to rely on complementary logic gate signals to protect the inverter.

An effective protection against switch cross conduction is implemented by introducing commutation dead-times, i.e., suitable delays before the switch turn on signal is applied to the gate. The effect of dead-times is shown in Fig. 2.2, in the hypothesis that a positive current I_O is flowing through the load. The figure assumes that a *period of observation* can be defined, whose duration is T_S, where switches S_1 and S_2 are meant to be on for times t_{ON1} and t_{ON2}, respectively, and where the load current is assumed to be constant (i.e., the load time constant L_S/R_S is assumed to be much longer than the observation period T_S). The existence of such an observation period guarantees that the definition of average load voltage is well posed. By that we simply mean the weighted average over time of the instantaneous load voltage in the period of observation.

To avoid cross conduction, the modulator delays S_1 turn-on by a time t_{dead}, applying the V_{GE1} and V_{GE2} command signals to the switches. The duration of t_{dead} is long enough to allow the safe turn-off of switch S_2 before switch S_1 is commanded to turn on, considering propagation delays through the driving circuitry, inherent switch turn off delays and suitable safety margins. The required dead-time duration is a direct function of the switch power rating and, for a typical 600 V, 40 A IGBT, is well below 1 µs.

Considering Fig. 2.2, it is important to notice that the effect of the dead-time application is the creation of a time interval where both switches are in the off state and the load current flows through the free-wheeling diodes. Because of that, a difference is produced between the desired duration of the S_1 switch on time and the actual one, that turns into an error in the voltage across the load. It is as well important to notice that the opposite commutation, i.e., where S_1 is turned off and S_2 is turned on, does not determine any such voltage error. However, we must point out that, if the load current polarity were reversed, the dead-time induced load voltage error would take place exactly during this commutation.

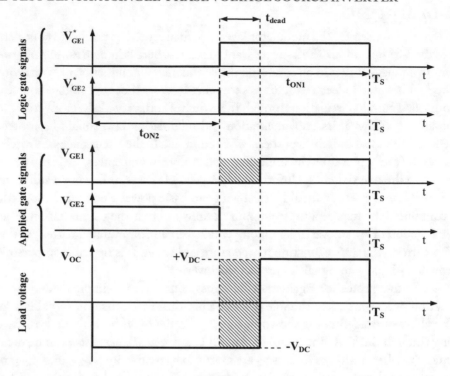

Figure 2.2: Dead-times effect: when a positive current I_O flows through the load, the actual on time for switch S_1 is shorter than the desired one. Consequently, the average voltage across the load is lower than the desired one.

The above discussion reveals that, because of dead-times, no matter what the modulator implementation is, an error on the load voltage will always be generated. This error ΔV_{OC}, whose entity is a direct function of dead-time duration and whose polarity depends on the load current sign according to the following relation

$$\Delta V_{OC} = -2V_{DC}\frac{t_{dead}}{T_S}\,\text{sign}\,(I_O), \qquad (2.2)$$

will have to be compensated by the current controller. Failure to do so will unavoidably determine a tracking error on the trajectory the load current has to follow (i.e., current waveform distortion). We will later see how some current controllers are inherently immune to dead-time induced distortion, while others are not.

We cannot end this discussion of dead-times without adding that, motivated by the considerations above, several studies have been presented that deal with their *compensation*. Both off line, or feed-forward, techniques and closed-loop arrangements have been proposed to mitigate the problem. The interested reader can find very detailed discussions of these topics in technical papers like, for instance, [2] and [3].

2.2 LOW-LEVEL CONTROL OF THE VOLTAGE SOURCE INVERTER: PWM MODULATION

The definition of a suitable modulation law represents the first step in any converter control design. Several modulation techniques have been developed for switch mode power supplies: for the VSI case, the most successful is, undoubtedly, the pulse width modulation (PWM). Compared to other approaches, like pulse density modulation or pulse frequency modulation, the PWM offers significant advantages, for instance in terms of ease of implementation, constant frequency inverter operation, immediate demodulation by means of simple low-pass filters. The analog implementation of PWM, also known as *naturally sampled* PWM, is indeed extremely easy, requiring, in principle, only the generation of a suitable carrier (typically a triangular or sawtooth waveform) and the use of an analog comparator. A simple PWM circuit is shown in Fig. 2.3.

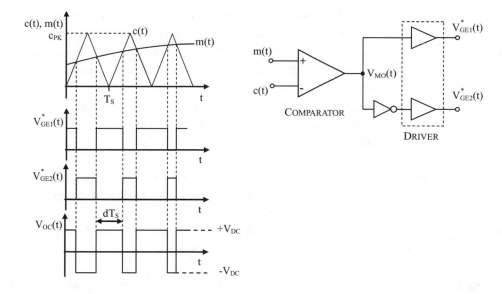

Figure 2.3: Analog implementation of a PWM modulator. The analog comparator determines the state of the switches by comparing the carrier signal $c(t)$ and the modulating signal $m(t)$. The figure shows the logic state of each switch and the resulting inverter voltage. No dead-time is considered.

2.2.1   ANALOG PWM: THE NATURALLY SAMPLED IMPLEMENTATION

Considering the circuit and what has been explained in Section 2.1, it is easy to see that, as a result of the analog comparator and driving circuitry operation, a square-wave voltage V_{OC} will be applied to the load, with constant frequency $f_S = 1/T_S$, T_S being the period of the carrier signal $c(t)$, and variable duty-cycle d. This is implicitly defined, again from Fig. 2.3, as the ratio between the time duration of the $+V_{DC}$ voltage application period and the duration of the whole modulation period, T_S. Finally, Fig. 2.3 allows to see the relation between duty-cycle and the average value (in the modulation period) of the load voltage, that is calculated in the Aside 1.

It is now interesting to explicitly relate signal $m(t)$ to the resulting PWM duty-cycle. Simple calculations show that, in each modulation period, where a constant m is assumed, the following equation holds:

$$\frac{m}{d\,T_S} = \frac{c_{PK}}{T_S} \quad \Leftrightarrow \quad d = \frac{m}{c_{PK}}. \tag{2.3}$$

If we now assume that the modulating signal changes slowly along time, with respect to the carrier signal, i.e., that the upper limit of $m(t)$ bandwidth is well below $1/T_S$, we can still consider the result (2.3) correct. This means that, in the hypothesis of a limited bandwidth $m(t)$, the information carried by this signal is transferred by the PWM process to the duty-cycle, that will change slowly along time following the $m(t)$ evolution. The duty-cycle, in turn, is transferred to the load voltage waveform by the power converter. The slow variations of the load voltage average value will therefore copy those of signal $m(t)$.

The simplified discussion above may be replaced by a more mathematically sound approach, that an interested reader can find in power electronics textbooks like [4], [5], and [6]. However, this approach would basically show that the harmonic frequency content, i.e., the spectrum, of the modulating signal $m(t)$ is shifted along frequency by the PWM process, and is replicated around all integer multiples of the carrier frequency. This implies that, as long as the spectrum of signal $m(t)$ has a limited bandwidth with a upper limit well below half of the carrier frequency, signal demodulation, i.e., the reconstruction of signal $m(t)$ spectrum from the signal $V_{OC}(t)$, with associated power amplification, can be easily achieved by low-pass filtering $V_{OC}(t)$. In the case of power converters, like the one we are considering here, the low-pass filter is actually represented by the load itself.

Referring again to Fig. 2.1 and to Aside 1, it is possible to see that the transfer function between the inverter voltage V_{OC} and load current I_O indeed presents a single-pole low-pass filter frequency response. The pole is located at an angular frequency that is equal to the ratio between load resistance R_S and load inductance L_S. Because of that, we can assume that, if the load *time constant*, L_S/R_S, is designed to be much higher than the modulation period T_S, the load current I_O average in the modulation period will precisely follow the trajectory determined by signal $m(t)$. This is the situation described by Fig. 2.4. It is worth noting that, while the average current is suitably sinusoidal, the instantaneous current waveform is characterized by a residual switching noise, the current *ripple*. This is a side effect determined by the non ideal filtering of high-order modulation harmonics, given by the load low-pass characteristics.

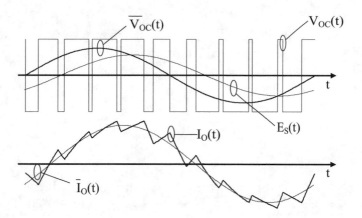

Figure 2.4: Example of PWM application to the VSI of Fig. 2.1. The instantaneous load voltage $V_{OC}(t)$ is demodulated by the low-pass filter action of the inverter load. The resulting load current $I_O(t)$ has an average value, $\bar{I}_O(t)$ whose waveform is determined by the instantaneous voltage average value $\overline{V}_{OC}(t)$ (and by load voltage E_S, here assumed to be sinusoidal).

Aside 1. VSI State Space Model

The VSI represented in Figure 2.1 can be described in the state space by the following equations:

$$\begin{cases} \dot{x} = Ax + Bu \\ y = Cx + Du \end{cases}, \tag{A1.1}$$

where $x = [I_O]$ is the state vector, $u = [V_{OC}, E_S]^T$ is the input vector, and $y = [I_O]$ is the output variable. In this very simple case, the state and output vectors have unity size, but, in the general case, higher sizes can be required to correctly model the converter and its load. Direct circuit inspection yields

$$A = [-R_S/L_S], \quad B = [1/L_S, -1/L_S], \quad C = [1], \quad D = [0, 0]. \tag{A1.2}$$

Based on this model and using Laplace transformation, the transfer function between the inverter voltage V_{OC} and the output current I_O, $G_{I_O V_{OC}}$ can be found to be

$$G_{I_O V_{OC}}(s) = C \cdot (sI - A)^{-1} \cdot B_{11} = \frac{1}{R_S} \cdot \frac{1}{1 + s\dfrac{L_S}{R_S}}. \qquad (A1.3)$$

The transfer function (A1.3) relates variations of the inverter voltage V_{OC} to the consequent variations of the output current I_O. The relation has been derived under no restrictive hypothesis, meaning that it has a general validity. In particular, (A1.3) can be used to relate variations of the *average* values of V_{OC} and I_O, where by average of any given variable v we mean the following quantity:

$$\bar{v}(t) = \frac{1}{T_S} \int_t^{t+T_s} v(\tau) \, d\tau, \qquad (A1.4)$$

where T_S is our observation and averaging interval. In the particular case of PWM control, the definition (A1.4) is well posed once the averaging period T_S is taken equal to the modulation period.

Considering now the input variable V_{OC}, we can immediately calculate its average value as a function of the PWM duty-cycle. This turns out to be equal to

$$
\begin{aligned}
\overline{V}_{OC}(t) &= \frac{1}{T_s} \int_t^{t+T_s} V_{OC}(\tau) \, d\tau \\
&= \frac{1}{T_S} (T_S \cdot V_{DC} \cdot d(t) - V_{DC}(1 - d(t)) \cdot T_S) = V_{DC}(2d(t) - 1), \qquad (A1.5)
\end{aligned}
$$

where $d(t)$ is the duty-cycle, as defined in Section 2.2. We can now easily calculate the relation between variations of the duty-cycle d and variation of $\overline{V}_{OC}$. Perturbation of (A1.5) yields

$$\frac{\partial \overline{V}_{OC}}{\partial d} = 2V_{DC}, \qquad (A1.6)$$

where V_{DC} is assumed to be constant. In the assumption of small perturbations around any given operating point, the transfer function between duty-cycle and load current can be obtained substituting (A1.6) into (A1.3). We find

$$G(s) = \frac{\tilde{I}_O(s)}{\tilde{D}(s)} = \frac{2V_{DC}}{R_S} \cdot \frac{1}{1 + s\dfrac{L_S}{R_S}}, \qquad (A1.7)$$

where $\tilde{I}_O(s)$ and $\tilde{D}(s)$ represent the Laplace transformed small perturbations of the variables I_O and d around any selected operating point. The result (A1.7) can be used in the design of current regulators.

In general, we will see how the removal of such switching noise from the control signals, that is essential for the proper operation of any digital controller, is fairly easy to achieve, even without using further low pass filters in the control loop.

In the following sections, we will see how a current controller can be designed. The purpose of the current controller will be to automatically generate signal $m(t)$ based on the desired load current trajectory, that will be designated as the *current reference* signal.

Before we move to digital PWM and current control design, there is a final issue to consider, related to the dynamic response of the PWM modulator [7, 8, 9, 10, 11]. Considering the circuit in Fig. 2.3, it is possible to see that a sudden change in the modulating signal amplitude always implies an immediate, i.e., within the current modulation period, adjustment of the resulting duty-cycle. This means that the analog implementation of PWM guarantees the minimum delay between modulating signal and duty-cycle. This intuitive representation of the modulator operation can be actually corroborated by a more formal, mathematical analysis. Indeed, the derivation of an equivalent modulator transfer function, in magnitude and phase, has been studied and obtained since the early 1980's. The modulator transfer function has been determined using small-signal approximations [7], where the modulating signal $m(t)$ is decomposed in a DC component M and a small-signal perturbation $\tilde{m}$ (i.e., $m(t) = M + \tilde{m}(t)$). Under these assumptions, in [7], the author demonstrates that the phase lag of the naturally sampled modulator is actually zero, concluding that the analog PWM modulator delay can always be considered negligible. Quite differently, we will see in the following section how the discrete time or digital implementations of the pulse width modulator [8], that necessarily imply the introduction of sample-and-hold effects, determine an appreciable, not at all negligible, delay effect.

2.2.2   DIGITAL PWM: THE UNIFORMLY SAMPLED IMPLEMENTATION

The basic principles described in Section 2.2.1 apply also to the digital implementation of the PWM modulator. In the more direct implementation, also known as "uniformly sampled PWM," each analog block is replaced by a digital one. The analog comparator function is replaced by a digital comparator, the carrier generator is replaced by a binary counter and so forth. We can see the typical hardware organization of a digital PWM, of the type we can find inside several microcontrollers and digital signal processors, either as a dedicated peripheral unit or as a special, programmable function of the general purpose timer, in Fig. 2.5.

The principle of operation is straightforward: the counter is incremented at every clock pulse; any time the binary counter value is equal to the programmed duty-cycle (match condition), the binary comparator sets the gate signal low and, if so programmed, triggers an interrupt to the microprocessor. The gate signal is set high at the beginning of each counting (i.e., modulation) period, where another interrupt is typically generated for synchronization purposes. The counter and comparator have a given number of bits, n, which is often 16, but can be as low as 8, when a very simple microcontroller is used. Actually, depending on the ratio between the durations of the modulation period and the counter clock period, a lower number of bits, N_e, could be

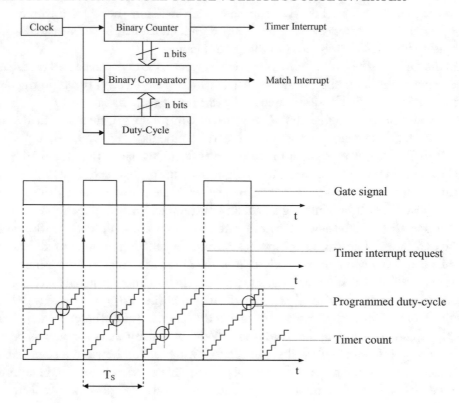

Figure 2.5: Simplified organization of a digital pulse width modulator. At the beginning of the counting period, the gate signal is set to high and goes low at the *match condition* occurrence, i.e., any time the binary counter value is equal to the programmed duty-cycle. The binary counter triggers an interrupt request for the microprocessor at the beginning of each modulation period.

available to represent the duty-cycle. The N_e parameter is also important to determine the duty-cycle quantization step, that can have a significant impact on the generation of limit cycles, as we will explain in the following chapter. For now it is enough to say that, with this type of modulator, the number N_e of bits needed to represent the duty-cycle is given by the following relation:

$$N_e = \text{floor}\left[\frac{\log_{10}\left(\dfrac{f_{clock}}{f_S}\right)}{\log_{10} 2}\right] + 1, \tag{2.4}$$

where f_{clock} is the modulator clock frequency, $f_S = 1/T_S$ is the desired modulation frequency and the *floor* function calculates the integer part of its argument. Typical maximum values for

f_{clock} are in the few tens of MHz range, while modulation frequencies can be as high as a few hundreds of kHz. Therefore, when the desired modulation period is short, the number of bits, N_e, given by (2.4) will be much lower than the number bits, n, available in the comparator and counter circuits, unless a very high clock frequency is possible.

Figure 2.5 allows us to discuss another interesting issue about digital PWM, that is the dynamic response delay of the modulator. In the considered case, it is immediate to see that the modulating signal update is performed only at the beginning of each modulation period. We can model this mode of operation as a *sample and hold* effect. We can observe that, if we neglect the digital counter and binary comparator operation assuming infinite resolution, the digital modulator works exactly as an analog one, where the modulating signal $m(t)$ is sampled at the beginning of each modulation period and the sampled value held constant for the whole period.

It is now evident that, because of the sample and hold effect, the response of the modulator to any disturbance, e.g., to one requiring a step change in the programmed duty-cycle value, can take place only during the modulation period *following* the one where the disturbance actually takes place. Please note that this delay effect amounts to a dramatic difference with respect to the analog modulator implementation, where the response could take place already during the *current* modulation period, i.e., with negligible delay. Therefore, even if our signal processing were fully analog, without any calculation or sampling delay, passing from an analog to a digital PWM implementation would imply an increase in the system response delay. We will see how this simple fact implies a significant reduction of the system's phase margin with respect to the analog case, that often compels the designer to adopt a more conservative regulator design and to accept a lower-closed loop system bandwidth.

Since these issues can be considered fundamental for all the following discussions, from the intuitive considerations reported above, we can now move to a precise small-signal Laplace-domain analysis, that might be very useful for a clear understanding of control limitations and delay effects implied by the uniformly sampled PWM.

An equivalent model of the uniformly sampled PWM process is represented in Fig. 2.6(a). As can be seen, the schematic diagram adopts the typical continuous time model of a sampled data system, where an ideal sampler is followed by a zero-order-hold (ZOH). The quantization effect that is associated, in the physical implementation of the modulator of Fig. 2.5, to the digital counter and binary comparator operation, is neglected, being irrelevant from the dynamic response delay standpoint. Accordingly, in the model of Fig. 2.6(a), after the modulating signal $m(t)$ is processed by the ZOH, the PWM waveform is generated by an ideal analog comparator, that compares the ZOH output signal $m_s(t)$ and the carrier waveform $c(t)$.

Depending on $c(t)$, several different uniformly-sampled pulse-width modulators can be obtained. For example, in Fig. 2.6(b) a trailing-edge modulation is depicted, where the update of the modulating signal is performed at the beginning of the modulation period. Note that this is an exactly equivalent representation of the modulator organization of Fig. 2.5. In a small-signal

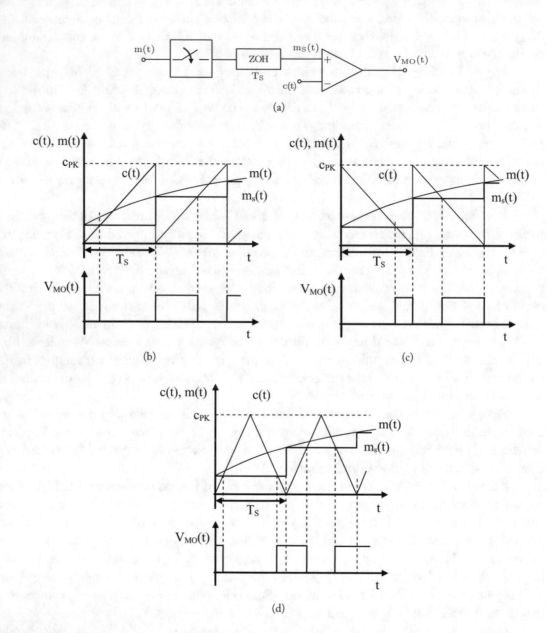

Figure 2.6: Uniformly sampled PWM with single update mode: (a) general block diagram, (b) trailing-edge modulation, (c) leading-edge modulation, and (d) triangular carrier modulation.

approximation, it is possible to find that the transfer function between the modulating signal $m(t)$ and the output of the comparator $V_{MO}(t)$ is given by [7]:

$$\text{PWM}(s) = \frac{V_{MO}(s)}{M(s)} = \frac{e^{-sDT_S}}{c_{PK}}, \qquad (2.5)$$

where $V_{MO}(s)$ and $M(s)$ represent the Laplace transforms of $V_{MO}(t)$ and $m(t)$, respectively. Therefore, the uniformly sampled modulator presents a delay whose value is proportional to the steady-state duty-cycle D. In more general terms, the delay introduced by the PWM modulator can be defined as the time distance between the modulating signal $m(t)$ sampling instant and the instant when the output pulse is completely determined (i.e., when $m_s(t)$ intersects $c(t)$ in Fig. 2.6). Based on this definition, the result (2.5) has been extended also to other types of modulator organizations (trailing edge, triangular carrier, etc.) [8]. For example, for the leading-edge modulation represented in Fig. 2.6(c), the small-signal modulator transfer function turns out to be:

$$\text{PWM}(s) = \frac{V_{MO}(s)}{M(s)} = \frac{e^{-s(1-D)T_S}}{c_{PK}} \qquad (2.6)$$

while, for the triangular carrier modulation, where the sampling of the modulating signal is done in the middle of the switch *on* period (Fig. 2.6(d)), it is:

$$\text{PWM}(s) = \frac{V_{MO}(s)}{M(s)} = \frac{1}{2c_{PK}} \left(e^{-s(1-D)\frac{T_S}{2}} + e^{-s(1+D)\frac{T_S}{2}} \right). \qquad (2.7)$$

Finally, the case of the triangular carrier modulator, where the sampling of the modulating signal is done in the middle of the switch *off* period, can be simply derived from (2.7) substituting D with D', being $D' = 1 - D$.

2.2.3 SINGLE UPDATE AND DOUBLE UPDATE PWM MODES

To partially compensate for the increased delay of the uniformly sampled PWM, the double update mode of operation is often available in several microcontrollers and DSPs. In this mode, the duty-cycle update is allowed at the beginning and at the half of the modulation period. Consequently, in each modulation period, the match condition between counter and duty-cycle registers is checked twice, at first during the run-up phase, then during the run-down phase.

In the occurrence of a match, the state of the gate signal is toggled. As can be seen in Fig. 2.7, the result of this mode of operation is a stream of gate pulses that are symmetrically allocated within the modulation period, at least in the absence of any perturbation. Interrupt requests are generated by the timer at the beginning and at the half of the modulation period, to allow proper synchronization with other control functions, e.g., with the sampling process.

It is also evident from Fig. 2.7 that, in the occurrence of a perturbation, the modulator response delay is reduced, with respect to the single update case because, now, the duty-cycle update can be performed at the occurrence of each half period interrupt request. In this case,

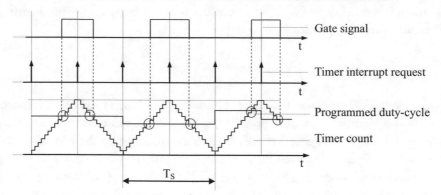

Figure 2.7: Double update mode of operation for a digital pulse width modulator. Duty-cycle update is allowed at the beginning and at a half of the modulation period. Note that the gate pulses are now symmetrically allocated within the modulation period (in the steady state).

however, an asymmetric pulse is generated, but symmetry is restored immediately afterwards, so that its temporary loss is of little consequence.

Maybe less evident is the drawback of this operating mode: given the number of bits, N_e, needed to represent the duty-cycle and the clock frequency f_{clock}, the switching period has to be doubled to contain both the run up and run down phases. Of course, it is possible to maintain the same modulation frequency of the single update case, but, in order to do that, either the clock frequency needs to be doubled or the number of bits needs to be reduced by one.

Following the reasoning reported in the previous section, we can derive an exact, continuous time equivalent model also of the digital PWM with double duty-cycle update. A representation of this model is shown in Fig. 2.8. Simple calculations show that the small-signal modulator transfer function is, in this case, given by [8]:

$$\text{PWM}(s) = \frac{V_{MO}(s)}{M(s)} = \frac{1}{2c_{PK}}\left(e^{-sD\frac{T_S}{2}} + e^{-s(1-D)\frac{T_S}{2}}\right). \tag{2.8}$$

It is interesting to compare the modulator phase lag for the single and double update modes of operation. In (2.7), we find $arg(\text{PWM}(j\omega)) = -\omega T_S/2$ while, in (2.8), $arg(\text{PWM}(j\omega)) = -\omega T_S/4$ so that, as it could be expected, the modulator phase lag is reduced by one half in the double update mode. This property can give significant benefits, in terms of the achievable speed of response, for any controller built on top of the digital modulator.

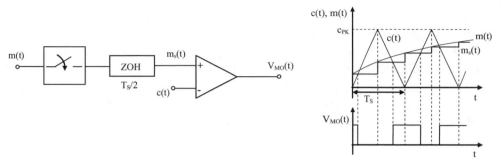

Figure 2.8: Model of the uniformly sampled PWM with double update.

2.2.4 MINIMIZATION OF MODULATOR DELAY: A STRONG MOTIVATION FOR MULTI-SAMPLING

In the most recent studies concerning digital control of power converters the key role played by the modulator delay in limiting the achievable control bandwidth has been very well clarified. A different approach has been suggested, that exploits the possibility of sampling control variables, and consequently adjusting the duty-cycle, several times (e.g., 4, 8, 16 times) within the modulation period. The purpose of this is to reduce the PWM response delay and increase the system phase margin, extending the benefits seen for the double update in comparison with the single update mode.

In order to evaluate the modulator phase lag, let us consider the system shown in Fig. 2.9: the modulating signal is sampled N times during the switching period, so that the sampling time interval is now $T_{sampling} = T_S/N$; moreover, in order to fully exploit the advantages of the multiple sampling technique, the control algorithm updates the control signal $m(t)$ at each sampling event. In the multi-sampled case, the PWM is modeled with an equivalent system similar to that shown in Fig. 2.6, with the only difference that the input signal $m_s(t)$ is now a sequence of variable amplitude pulses, updated with frequency $f_{sampling} = N \cdot f_S$. Accordingly, the hold time of the ZOH is now $T_{hold} = T_{sampling} = T_S/N$. It can be shown that the low-frequency, small-signal behavior of the multi-sampled digital PWM is again that of a pure delay,

$$\text{PWM}(s) = \frac{1}{c_{PK}} e^{-st_d}, \qquad (2.9)$$

but the equivalent delay time is now given by:

$$t_d = DT_S - \frac{floor[N \cdot D]}{N} T_S, \qquad (2.10)$$

where $floor[N \cdot D]$, as usual, yields the integer part of the fractional number $N \cdot D$. Equations (2.9) and (2.10) can be derived analytically with methods similar to those used in [7], for the uniformly sampled modulator, and applying a small-signal approximation. The first term $D \cdot T_S$ in (2.10) is the same delay found in (2.5), and does not depend on the multi-sampling factor N. The second term takes into account the multiple sampling effect, which is primarily that of reducing the equivalent delay time, and thus the total phase lag introduced by the PWM. Moreover, from (2.10) we can infer that, as N tends to infinity, the equivalent delay time tends to zero. The result is obvious, since when N is high the multi-sampled PWM approaches the naturally sampled modulator, where the phase lag is known to be zero.

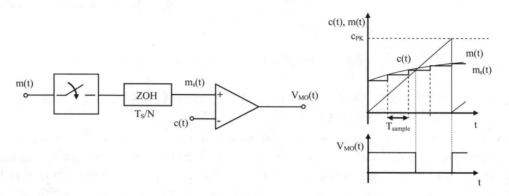

Figure 2.9: Multi-sampled PWM.

The main drawback of such an approach is represented by the need for proper filtering of the switching noise from the control signals, that is, instead, straightforward with the single or double update mode. Filtering the control signals may impair the system phase margin, reducing the advantage of the multi-sampled strategy. We will come back to this issue in Chapters 3 and 4, where we will open the discussion of digital controllers and treat the multi-sampled implementations in more detail. For now, it will be enough to say that some research is in progress around the world to find means to get the needed filtering without worsening the system stability margin, for example using sophisticated estimation techniques or particular controller organizations. The last remark about multi-sampling refers to the hardware required for the implementation. This is significantly different from what can be considered the standard PWM organization, available with off the shelf microcontrollers and DSPs, and calls for other solutions, e.g., the use of hardware programmable digital control circuits, like those based on Field Programmable Gate Arrays (FPGAs).

2.3   ANALOG CONTROL APPROACHES

We begin here to deal with the control problem this book is all about. In order to better appreciate the merits and limitations of the digital approach, we will now briefly discuss two possible analog implementations of a current control loop: the PI linear controller and the non linear hysteresis controller. We refer to our test case, as represented in Fig. 2.1, but in order to make some explicit calculations, we will take into account the parameters listed in Table 2.1.

Table 2.1: Half-bridge inverter parameters

Rated output power, P_O	1500 (W)
Phase inductance, L_S	1.2 (mH)
Phase resistance, R_S	1 (Ω)
Phase voltage, E_S	120 (V_{RMS})
Load frequency, f_O	60 (Hz)
DC link voltage, V_{DC}	250 (V)
Switching frequency, f_S	20 (kHz)
PWM carrier peak, c_{PK}	4 (V)
Current transducer gain, G_{TI}	0.1 (VA^{-1})

In this example we suppose the purpose of the VSI is to deliver a given amount of output power P_O to the load, that is represented by the voltage source E_S. Resistor R_S may represent the lossy elements of the load and of the inverter inductor. What we are discussing can be thought as the typical grid-tied inverter application, where a sinusoidal current of suitable amplitude and frequency equal to the E_S voltage frequency f_O, must be generated. Consequently, we have also taken into account the presence of a current transducer, whose gain, G_{TI}, is given in Table 2.1, and that may be in practice implemented by a Hall sensor.

For the controller implementation, we can assume that one of the average current mode control integrated circuits available on the market is used. This will generally include all the needed functions, from error amplification and loop compensation to PWM modulation. Of course, to keep the discussion simple, the presence of additional signal scale factors, for example due to internal voltage dividers, is not taken into account. Also, the PWM parameters reported in Table 2.1, although realistic, do not necessarily represent those of any particular integrated controller.

2.3.1 LINEAR CURRENT CONTROL: PI SOLUTION

Figure 2.10 shows the control loop block diagram, where all the components are represented by their respective transfer functions or gains. In particular, the controller block is represented by the typical proportional integral regulator structure, whose parameters K_P and K_I are going to be determined in the following. The output of the regulator represents the modulating signal that drives the pulse width modulator. This has been modeled as the cascade combination of two separate blocks: the first one is the modulator static gain, as given by (2.3), the second one is actually a first-order Padé approximation of its delay, considered equal to a half of the duration of the modulation period.

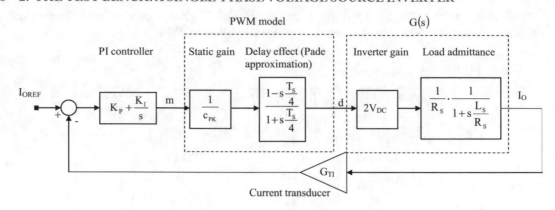

Figure 2.10: Control loop block diagram.

This choice deserves some clarification, since we have previously assessed the delay effect of an analog PWM to be negligible. The point is that, for reasons that will be fully motivated in Chapter 3, we are here considering the modulator as if it was *digitally* implemented, i.e., characterized by the sample and hold delay that we have previously described. From Section 2.2, we know that the equivalent model of the digital modulator can be given by (2.5), (2.6) or, possibly, (2.7). The proper characterization of these models is a little complicated. For this reason, in Fig. 2.10, we consider the response delay of the digital PWM to be, on average, equal to a half of the modulation period and we model this average delay with its first order Padé approximation. In Chapter 3, we will clearly account for this approximation and show that this is actually not penalizing.

Considering now the inverter and load models, we see that they are exactly based on the analysis presented in the Aside 1. Finally, to fully replicate a typical implementation, a transducer gain is taken into account. Additional filters, that are normally adopted to clean the transducer signal from residual switching noise, are instead not taken into account, in favor of a more essential presentation. Their transfer functions can be easily cascaded to the transducer block gain if needed. Given the block diagram of Fig. 2.10, the design of the PI compensator is straightforward. However, for the sake of completeness, we present the simple design procedure in the Aside 2.

Once the proper K_P and K_I values are determined, we still may want to check the system dynamic behavior and verify if a stable closed-loop controller with the desired speed of response has been obtained. In order to do that, before developing any converter prototype, it is very convenient to use one of the several dynamic system software simulators available. The simulation of the VSI depicted in Fig. 2.1, together with its current controller, gives the results described by Fig. 2.11. In particular, Fig. 2.11(a) shows the response of the closed loop system to a step change in the I_{OREF} current reference amplitude. It is possible to see that the closed loop plant

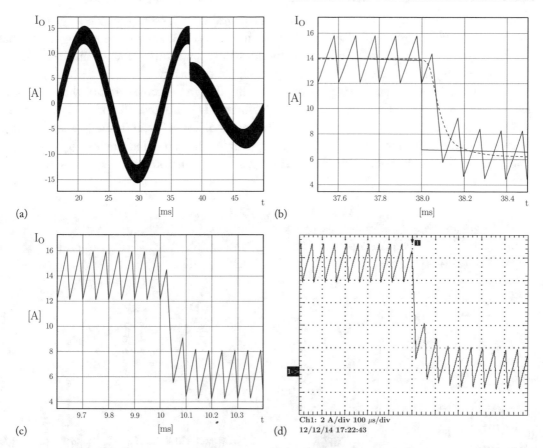

Figure 2.11: Simulation and experimental test of the VSI depicted in Fig. 2.1 with the controller designed according to the procedure reported in the Aside 2. The depicted variable is the VSI output current I_O. (a) Controller response to a step reference amplitude change. (b) Detail of previous figure. (c) PI controller step response in the experimental set-up conditions. (d) Experimental verification of the PI controller step response.

is properly controlled, with a sufficiently high phase margin not to incur in oscillations after the transient. Figure 2.11(b) shows the detail of the transient response: the controller reaches the new steady-state condition in three modulation periods, exhibiting no overshoots. It is worth noting that an anti wind-up action is included in the PI controller to prevent deep saturation of the integral controller during transients. One closing remark on Fig. 2.11(b) is due: an appreciable, albeit relatively small, steady-state tracking error between the reference signal (continuous line) and the instantaneous current average value (i.e., once the current ripple is filtered, dashed line), is visible both before and after the transient. This represents the residual tracking error of the current controller. As any other controller including an integral action, our PI is able to guarantee zero steady-state tracking error only for DC signals. In the case of an AC reference signal, as that of Fig. 2.11(b), a residual error will always be found, whose amplitude depends on the closed-loop system gain and phase at the particular reference signal frequency.

In order to verify the accuracy of the controller design procedure and of the adopted simulation models, the PI controller has been experimentally evaluated, thanks to a VSI hardware similar to the one of Fig. 2.1 and with the parameter set of Table 2.1. It is possible to observe how the step response transient Fig. 2.11(d) really resembles the simulated one of Fig. 2.11(c), with almost the same number of modulation periods required by the output current to settle onto the new reference level.

Aside 2. Design of the Analog PI Current Controller

At first, we want to determine the open loop gain for the block diagram of Fig. 2.10. This is given by the cascade connection of all blocks. We find

$$G_{OL}(s) = \left(K_P + \frac{K_I}{s} \right) \frac{2V_{DC}}{c_{PK}} \frac{1 - s\dfrac{T_S}{4}}{1 + s\dfrac{T_S}{4}} \frac{G_{TI}}{R_S} \frac{1}{1 + s\dfrac{L_S}{R_S}}. \tag{A2.1}$$

The regulator design is typically driven by specifications concerning the required closed-loop *speed of response* or, equivalently, the maximum allowed *tracking error* with respect to the reference signal. These specifications can be turned into equivalent specifications for the closed-loop bandwidth and phase margin. To give an example, we suppose that, for our current controller, a closed loop bandwidth, f_{CL}, equal to about one sixth of the switching frequency f_S is required, to be achieved with, at least, a 60° phase margin, ph_m.

We therefore have to determine the parameters K_P and K_I so as to guarantee the compliance to these requirements.

To rapidly get an estimation of the searched values, we suppose that we can approximate the open loop gain at the crossover angular frequency, i.e., at $\omega = \omega_{CL} = 2\pi f_{CL}$, with the following expression:

$$G_{OL}(j\omega_{CL}) \cong K_P \frac{2V_{DC}}{c_{PK}} \frac{1 - j\omega_{CL}\dfrac{T_S}{4}}{1 + j\omega_{CL}\dfrac{T_S}{4}} \frac{G_{TI}}{R_S} \frac{1}{1 + j\omega_{CL}\dfrac{L_S}{R_S}}, \tag{A2.2}$$

which, in principle, will be a good approximation as long as $K_I \ll \omega_{CL} K_P$ (to be verified later). Imposing now the magnitude of (A2.2) to be equal to one at the desired crossover frequency, we get

$$K_P = \frac{c_{PK}}{2V_{DC}} \frac{R_S}{G_{TI}} \sqrt{1 + \left(\omega_{CL}\frac{L_S}{R_S}\right)^2}. \tag{A2.3}$$

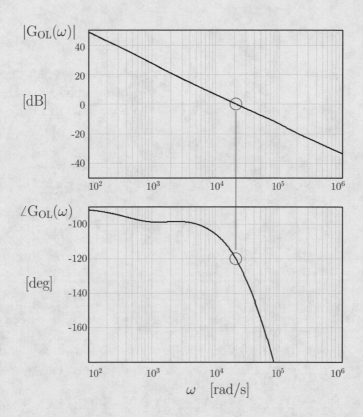

Figure A2.1: Bode plot of the open-loop gain.

The parameter K_I can then be calculated considering the open loop phase margin and imposing that to be equal to ph_m. We find from (A2.1)

$$-180° + ph_m = -90° - 2\tan^{-1}\left(\omega_{CL}\frac{T_S}{4}\right) - \tan^{-1}\left(\omega_{CL}\frac{L_S}{R_S}\right) + \tan^{-1}\left(\omega_{CL}\frac{K_P}{K_I}\right), \tag{A2.4}$$

which yields

$$K_I = \frac{\omega_{CL}K_P}{\tan\left(-90° + ph_m + 2\tan^{-1}\left(\omega_{CL}\frac{T_S}{4}\right) + \tan^{-1}\left(\omega_{CL}\frac{L_S}{R_S}\right)\right)}. \tag{A2.5}$$

Note that (A2.5) is exact; only the K_P value is obtained through an approximation. Considering the parameters listed in Table 2.1 and $\omega_{CL} = 2\pi \, f_S/6 \cong 20.94 \, \text{krad s}^{-1}$, we can immediately find the following values:

$$K_P = 2.012$$
$$K_I = 2.162 \times 10^3 \; (\text{rad s}^{-1}).$$

It is easy to verify that the condition $K_I \ll \omega_{CL} K_P$ is reasonably met by this solution. Nevertheless, in order to explicitly evaluate the quality of the approximated solution, we can compare the values above with the solutions of the *exact* design equations. We practically need to solve the following system of equations:

$$
\begin{cases}
\dfrac{K_I}{K_P} = \dfrac{\omega_{CL}}{\tan\left(-90° + ph_m + 2\tan^{-1}\left(\omega_{CL}\dfrac{T_S}{4}\right) + \tan^{-1}\left(\omega_{CL}\dfrac{L_S}{R_S}\right)\right)} \\[4ex]
K_P = \dfrac{c_{PK}}{2V_{DC}}\dfrac{R_S}{G_{TI}}\sqrt{\dfrac{1 + \left(\omega_{CL}\dfrac{L_S}{R_S}\right)^2}{1 + \left(\dfrac{1}{\omega_{CL}}\dfrac{K_I}{K_P}\right)^2}}
\end{cases}
\tag{A2.6}
$$

The solution yields $K_P = 2.01$, $K_I = 2.159 \times 10^3 \; (\text{rad s}^{-1})$.

As can be seen, the exact values are very close to those found by the approximated procedure above. This happens in the large majority of practical cases, so that (A2.3) and (A2.5) can be very often directly used.

As a final check of the design, in Fig. A2.1 we present the Bode plot of the open-loop gain, where the desired crossover frequency and phase margin can be read.

An interesting advantage of the PI current controller usage is the automatic compensation of dead-time induced current distortion. Referring to our brief discussion of Section 2.1.4, it is possible to see how, from the current controller standpoint, the dead-time effect can be equivalently seen as a disturbance signal that sums with the average inverter output voltage, generated by an ideal (i.e., with no dead times) pulse width modulator. If the dead-time duration can be considered constant, as is often the case, the disturbance signal is very close to a square wave, whose amplitude is directly proportional to the dc link voltage and to the dead time duration and inversely proportional to the switching period duration (2.2). Compared to the output current signal, this square wave has the same frequency and *opposite* phase. We know that the PI controller guarantees a significantly higher than unity open loop gain at the current reference frequency (see the Bode plot in the Aside 2), which is typically maintained for at least a decade above. As a result, the controller will reject the disturbance quite effectively: only minor crossover effects, due to an incomplete compensation of the higher order harmonics of the square wave, will be observable on the output current waveform.

2.3.2 NONLINEAR CURRENT CONTROL: HYSTERESIS CONTROL

The PI controller discussed above is not the only possible solution to provide the VSI of Fig. 2.1 with a closed loop current control. Other approaches are viable, among which the hysteresis current controller is one of the most successful. Even if we are not going to develop this topic in detail, we still would like to briefly describe the principles of this type of analog current controller, just not to give to the reader the wrong feeling that analog current control only amounts to PI regulators and PWM.

It is important to underline from the start that the hysteresis controller is a particular type of bang-bang non linear control and, as such, the dynamic response it is able to guarantee is extremely fast; actually it is the fastest possible for any VSI with given DC link voltage and output inductance. The basic reason for this is that the hysteresis controller does not require any modulator: the state of the converter switches is determined directly by comparing the instantaneous converter current with its reference. A typical hysteresis current controller is depicted in Fig. 2.12.

As can be seen, an analog comparator is fed by the instantaneous current error, and its output directly drives the converter switches. Thanks to the VSI topology and to the fact that the DC link voltage V_{DC} will always be higher than the output voltage E_S peak value, the current derivative sign will be positive any time the high side switch is closed and negative any time the low side switch is closed. This guarantees that the controller organization of Fig. 2.12 will maintain the converter output current always close to its reference, the maximum instantaneous error being limited to the hysteresis band width. At the limit condition of zero hysteresis band width, the current error can be forced to zero as well: unfortunately, this condition implies an infinite frequency for the switch commutations, which is, of course, not practical. In real life implementations, the hysteresis band width is kept sufficiently small to minimize the tracking error without implying too high switching frequencies. As a consequence, also the compensation of dead-time induced current distortion will be very good.

What is even more important, in case of any transient, that may bring the instantaneous current outside the hysteresis band, the controller will almost immediately close the right switch to bring the current back inside the band, thus minimizing the response delay and tracking error. An example of the excellent large-signal speed of response is shown in Fig. 2.13. Clearly, there is no linear controller that can be faster than this.

Nevertheless, the hysteresis current controller is not ubiquitously used in power electronics. That's because, despite its speed of response and high quality reference tracking capabilities, this type of controller does have some drawbacks as well. The main one is represented by a variable switching frequency. Indeed, any time the current reference is not constant the converter switching frequency will vary along the current reference period. The same holds in the case the output

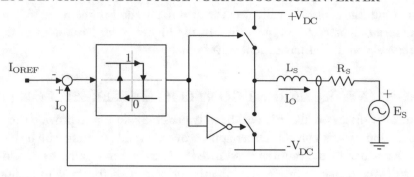

Figure 2.12: Hysteresis current control hardware organization.

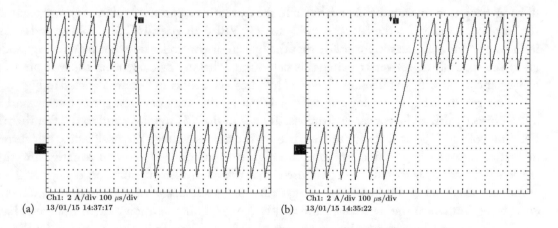

Figure 2.13: Experimental test of the hysteresis current controller step response: (a) -10 A step transient; and (b) $+10$ A step transient. Measured switching frequency $f_S = 18$ kHz.

voltage E_S is variable. The range of frequency variation can be very large, thus making the proper filtering of the high frequency components of voltages and currents quite expensive. Moreover, in the VSI applications like controlled rectifiers or active filters, the injection of a variable frequency noise into the utility grid is not recommended, because unpredictable resonances with other connected loads could be triggered. To solve this and other problems a considerable research activity has been developing in the last few years. Different control solutions, that try to keep the benefits of the hysteresis controller and, for example, get a fixed switching frequency out of it, have been proposed. We are going to deal with these more advanced topics in Chapter 4 where a fully digital, fixed frequency implementation of the hysteresis controller will be discussed. Besides, the interested reader can find useful information in several technical papers like [12] or [13].

REFERENCES

[1] ISOSMART™ Half Bridge Driver Chipset, IXBD4410/4411 Data sheet and Application note, © 2004, IXYS website. 9

[2] N. Urasaki, T. Senjyu, K. Uezato,T. Funabashi, "An Adaptive Dead-Time Compensation Strategy for Voltage Source Inverter Fed Motor Drives," *IEEE Transactions on Power Electronics,* Vol. 20, No. 5, September 2005, pp. 1150–1160. DOI: 10.1109/TPEL.2005.854046. 13

[3] A.R. Munoz, T.A. Lipo, "On-Line Dead-Time Compensation Technique for Open-Loop PWM-VSI Drives," *IEEE Transactions on Power Electronics,* Vol. 14, No. 4, July 1999, pp. 683–689. DOI: 10.1109/63.774205. 13

[4] N. Mohan. T. Undeland, W. Robbins, "Power Electronics: Converters, Applications and Design," 2003, Wiley, ISBN 0–471-22693-9. 14

[5] J. Kassakian, G. Verghese, M. Schlecht, "Principles of Power Electronics," 1991, Addison Wesley, ISBN 02010–9689-7. 14

[6] R. W. Erickson, D. Maksimovic, "Fundamentals of Power Electronics," Second Edition, 2001, Springer, ISBN 07923–7270-0. DOI: 10.1007/b100747. 14

[7] R.D. Middlebrook; "Predicting modulator phase lag in PWM converter feedback loops," Advances in switched-mode power conversion, vol I, pp. 245–250, 1981. 17, 21, 23

[8] D.M. Van de Sype, K. De Gusseme, A.P. Van den Bossche, J.A. Melkebeek, "Small-signal Laplace-domain analysis of uniformly-sampled pulse-width modulators;" Power Electronics Specialists Conference (PESC), 20–25 June, 2004, pp. 4292–4298. DOI: 10.1109/PESC.2004.1354760. 17, 21, 22

[9] D.M. Van de Sype, K. De Gusseme, A.R. Van den Bossche, J.A. Melkebeek, "Small-signal z-domain analysis of digitally controlled converters," Power Electronics Specialists Conference (PESC), 20–25 June, 2004, pp. 4299–4305. DOI: 10.1109/PESC.2004.1354761. 17

[10] G.C. Verghese, M.E. Elbuluk, and J.G. Kassakian, "A general approach to sampled-data modeling for power electronic circuits," *IEEE Transactions on Power Electronics,* Vol. 1, No. 2, April 1986, pp. 76–89. DOI: 10.1109/TPEL.1986.4766286. 17

[11] G.R. Walker, "Digitally-Implemented Naturally Sampled PWM Suitable for Multilevel Converter Control," *IEEE Transactions on Power Electronics,* Vol. 18, No. 6, November 2003, pp. 1322–1329. DOI: 10.1109/TPEL.2003.818831. 17

[12] Q. Yao and D. G. Holmes, "A simple, novel method for variable-hysteresis-band current control of a three phase inverter with constant switching frequency," IEEE-IAS Annual Meeting, Toronto, ON, Canada, October 1993, pp. 1122–1129. DOI: 10.1109/IAS.1993.299038. 32

[13] S. Buso, S. Fasolo, L. Malesani, P. Mattavelli: "A Dead-Beat Adaptive Hysteresis Current Control," *IEEE Transactions on Industry Applications,* Vol. 36, No. 4, July/August 2000, pp. 1174–1180. DOI: 10.1109/28.855976. 32

CHAPTER 3

Digital Current Mode Control

In this chapter we begin the discussion of digital control techniques for switching power converters. In the previous chapter we have introduced the topology and operation of the half-bridge VSI and designed an analog PI current controller for this switching converter. Referring to that discussion, the first part of this chapter is dedicated to the derivation of a digital PI current controller resembling, as closely as possible, its analog counterpart. We will see how, by using proper *discretization* techniques, the continuous time design can be turned into a discrete time one, preserving, as much as possible, the closed loop properties of the former. It is important to underline from the beginning that the continuous time design followed by some discretization procedure is not the only design strategy we can adopt. *Discrete time design* is also possible, although its application is somewhat less common: as we will explain, its typical implementations rely on the use of *state feedback* and *pole placement* techniques. The second part of the chapter will describe in detail a remarkable example of discrete time design and, in doing that, it will also show how the synthesis of regulators that have no analog counterpart whatsoever can be implemented. This is the case of the *predictive* or *dead-beat* current controller.

3.1 REQUIREMENTS OF THE DIGITAL CONTROLLER

The first step in the design of a digital controller is always the implementation of a suitable *data acquisition* path. While signal acquisition organization is somehow *implicit* in analog control design, because both the plant and the controller operate in the *continuous* time domain, digital control requires particular care in signal conditioning and analog to digital conversion implementation. The reason for this is ultimately that, while the control signals are taken from a plant that operates in the continuous time domain, the operation of the controller takes place in the *discrete* time domain. Therefore, signals have to be converted from the continuous to the discrete time domain and, of course, the other way around. It is very important to be aware that not *every* implementation of this conversion process leads to a satisfactory controller performance. We will see how the control of conversion noise and the avoidance of *aliasing* phenomena play a critical role.

3.1.1 SIGNAL CONDITIONING AND SAMPLING

The typical organization of a digital current controller for the considered VSI is depicted in Fig. 3.1. Compared to Fig. 2.1, the power converter is represented here in a more compact form, using ideal switches and just a schematic representation of the driving circuitry, as these details are not essential for the following discussion. As can be seen, we assume the digital controller is developed using a microcontroller or digital signal processor (DSP) unit, with suitable built-in peripherals. Although this is not the only available option for the successful implementation of a digital controller, it is by far the most commonly encountered and the most straightforward to illustrate. Because of that, we will not discuss other possibilities, like the use of custom digital circuits or field programmable gate arrays (FPGAs), at this time, but rather come back to this interesting topic in Chapter 4.

Almost every microcontroller and several low-cost DSP units, typically identified as motion control DSPs or industrial application DSPs, include the peripheral circuits required by the set-up of Fig. 3.1. These are basically represented by an analog to digital converter (ADC) and a PWM unit. The data acquisition path for our current controller is quite simple; it just consists in the cascade connection of a current sensor, a properly designed signal conditioning electronic circuit and the ADC. It is worth adding some comments about the conditioning circuit, with respect to its general features described in Section 2.1.2, in order to relate its function more precisely to the ADC operation. From this point of view, the conditioning circuit has to guarantee that: (i) the sensor signal is amplified so as to fully exploit the input voltage range of the ADC and (ii) the signal is filtered so as to avoid *aliasing* effects.

The full exploitation of the ADC input voltage range is a key factor to reduce the quantization effects that may undermine control stability and/or reduce the quality of the regulation. The reason for this is that the number N_e of *effective* bits, that are used for the internal representation of the input signal samples, is maximum when the input voltage range is fully exploited. We can actually see that this number is given by the following relation

$$N_e = n - floor \left\lceil \frac{\log_{10} \frac{FSR}{S_{PP}}}{\log_{10} 2} \right\rceil, \qquad (3.1)$$

where S_{PP} is the peak to peak amplitude (in Volt) of the transduced input signal, FSR is the ADC full scale range (in Volt) and n is the ADC bit number. A little complication we typically find when designing the conditioning circuit is related to the input signal sign or polarity. It is quite common for the transduced current signal to be bipolar (i.e., to have both positive and negative sign), while the lower bound of the ADC voltage range is almost always zero. To take care of that, the conditioning circuit has to offset the input signal by a half of the ADC FSR. This operation associates the lower half of the ADC range to the negative values of the input signal, the upper half to the positive values. These simple considerations are normally enough to properly design the gain of the conditioning amplifier in the frequency band of interest. Given the expected peak to peak amplitude of the VSI output current and considering a suitable safety margin for the detection of overcurrent conditions, due to load transients or faults, it is immediately possible to determine the gain required to exploit the ADC full scale range.

The aliasing phenomenon is a consequence of the violation of Shannon's theorem, which defines the limitations for the exact reconstruction of a uniformly sampled signal [1]. The theo-

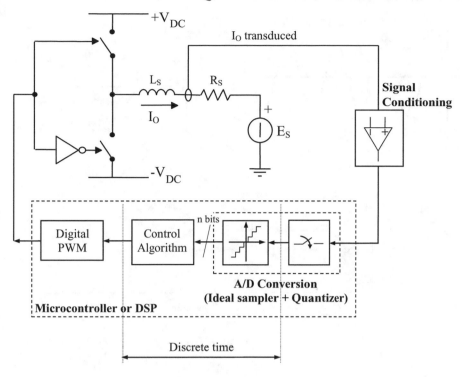

Figure 3.1: Typical organization of a digital current controller.

rem shows that there is an upper bound for the sampled signal bandwidth, beyond which perfect reconstruction, even by means of ideal interpolation filters, becomes impossible and aliasing phenomena appear. The limit frequency is called the Nyquist frequency, f_{Nyq}, and is proved to be equal to a half of the sampling frequency, $f_{sampling}$. In general, we will have to limit the frequency spectrum of the sampled signal by filtering, so as to make it negligible above the Nyquist frequency. This condition will determine the bandwidth and roll-off of the conditioning amplifier. A very intuitive graphical representation of the aliasing phenomenon is given in Fig. 3.2.

Another interesting issue, related to signal acquisition in digital control, is the definition of a suitable ADC model. From Fig. 3.1 we can see that the analog to digital conversion process can be mathematically modeled as the cascade connection of an ideal sampler and a n-bit uniform quantizer. The former is defined as a sampler whose output is a stream of *null duration* pulses, each having an amplitude equal to that of the input signal at the sampling instant. Its function is to model the actual sampling process, i.e., the transformation of the time variable from the continuous domain to the discrete domain, where time only exists as integer multiples of a fundamental unit, the sampling period. The latter is taken into account to model the loss of information implied by what can be interpreted as a *coding* procedure, where a continuous amplitude signal, i.e.,

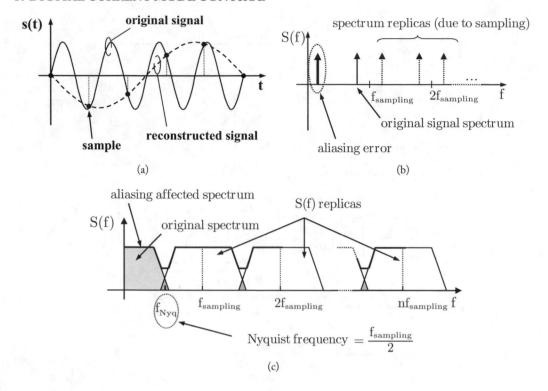

Figure 3.2: (a) Effect of a too low sampling frequency on the reconstructed signal (aliasing). (b) Interpretation of the aliasing effect of (a) in the frequency domain. Note how a low-frequency spectrum component is generated because of aliasing. (c) A more general situation: a distorted, aliasing-affected, spectrum is reconstructed because of the partial overlap of spectrum replicas.

a signal whose instantaneous level can vary with continuity in a given range of values, is transformed into a discrete amplitude signal, i.e., a digital signal, whose instantaneous level can only assume a finite number of values in the same given range. Because the possible discrete values can be interpreted as integer multiples of a fundamental unit, the quantization step Q, or, equivalently, the *least significant bit* LSB, the quantizer is called "uniform." Non uniform quantizers can sometimes be encountered, but very rarely in the kind of application we are interested in. For this reason, we will only discuss the uniform quantizer case. The typical transcharacteristic diagram for a uniform quantizer is shown in Fig. 3.3(a). As can be seen, a typical quantization noise e_q can be defined that is added to the signal as a result of analog to digital conversion. This can be interpreted as the loss of some of the information associated to the input signal, inherent to the analog to digital conversion and unavoidable. We will further discuss this phenomenon in one the following paragraphs. As far as the dynamic behavior of the ADC is concerned, it should be evident that both the quantizer and the ideal sampler are essentially instantaneous functions, that do not contribute to the dynamics of the system.

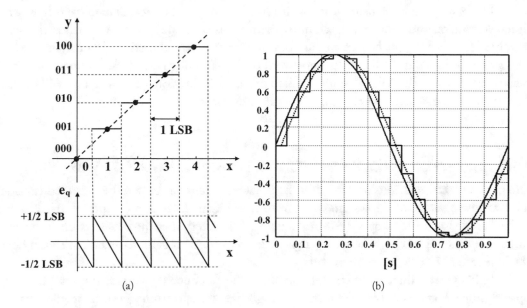

(a)   (b)

Figure 3.3: (a) Uniform quantizer trans-characteristic and quantization error e_q. (b) Sample and hold delay effect: compare the input signal (continuous line) and the reconstructed output signal, i.e., the fundamental harmonic component of the sampled signal (dotted line).

Figure 3.1 reveals another interesting point about the digital current controller organization, that is related to the digital PWM. This component processes the output of the control algorithm, a discrete time signal, and turns it into a continuous time signal, the state of the switches. This function, that represents the inverse of the sampling process and allows the controller to actuate the system under control, is known as *interpolation*. It is now evident that, from the digital control theory's standpoint, the DPWM is the part of our control system where interpolation takes place.

For reasons that will become clear in the following, it is often important to develop a *continuous time equivalent* model of the controller, i.e., of everything that is included between the sampler and the interpolator. In other words, we are often interested in a mathematical description of the digital controller as it is "seen" from the external, continuous time world's standpoint. This problem can be solved by considering what is known as a Zero Order Hold (ZOH) approximation of the interpolation process. Neglecting the presence of the control algorithm, we can describe this model simply considering that, in order to reconstruct the continuous time signal from the discrete time input samples, each sample value is held constant for the entire duration of the sampling period. It is actually possible to use different interpolation models [2], but, for the problems of our interest, this is normally a good enough one. We will see in the following how this approach is related to the DPWM equivalent continuous time models presented in Chapter 2.

However, it is immediate to recognize in this function a typical dynamic effect: any time a signal is sampled and converted again into a continuous time one by the interpolator, that we have now modeled as a simple holder, we cannot reconstruct exactly the original signal, but we have to face a delay effect that is directly proportional to the sampling period. An example of this effect is shown in Fig. 3.3(b). We will come back to this issue in Section 3.2.2, when we will discuss the digital controller design technique based on *discretization*.

3.1.2   SYNCHRONIZATION BETWEEN SAMPLING AND PWM

The general considerations presented in the previous paragraph have to be extended considering the particular nature of the system we want to control. As defined in Chapter 2, the VSI is controlled at the lowest level by a PWM modulator. This determines the presence, on each electrical variable, of the typical high frequency noise known as *ripple*. It is fundamental to clarify how this is taken care of in the sampling process.

It is evident that, in order not to violate the Shannon's theorem, the sampling process should proceed at a very high frequency, so high that the spectrum of the sampled signal might be considered negligible at the Nyquist frequency, even if a significant ripple is observable. This would require a sampling frequency at least one order of magnitude higher than the switching frequency. Unfortunately, hardware limitations do not allow the sampling frequency to become too high: we must keep in mind that our first controller implementation will be based on standard microcontroller or DSP hardware.

When we will discuss the adoption of multi-sampling strategies, we will see how they require a non conventional hardware organization, for example the use of FPGA circuits. In the typical case, instead, since the duty-cycle update is allowed at most twice per modulation period, in the double update mode of operation of the digital PWM, the sampling frequency cannot get higher than twice the switching frequency. Of course, in order to push the bandwidth of the closed loop plant as high as possible, we are normally not interested in sampling frequencies lower than the allowed maximum, at least for the current controller. When, in one of the following chapters, we will discuss the application of digital control to external loops, we will see how, sometimes, lower sampling frequencies can offer some advantages. However, in the case of the current controller, the sampling frequency should be maximized. The reason for this is quite obvious: by doing so the inherent sample and hold delay can be minimized and, consequently, the closed loop plant bandwidth can be maximized.

In conclusion, in the typical case, the sampling frequency will be set equal either to the switching frequency, or, if this is consistent with the available digital PWM implementation, to two times the switching frequency. But, if this is what we do, the Shannon's theorem conditions will always be violated!

This is one of the key issues in digital control applications to power electronic circuits: the typically recommended high ratio between sampling frequency and sampled signal bandwidth will never be possible. Nevertheless, we will shortly see how this, rather than detrimental, is normally advantageous for the controller effectiveness. The reason for this lies in *synchronization*.

If the sampling and switching processes are suitably synchronized, the effect of aliasing is the automatic reconstruction of the average value of the sampled signal, which is *exactly* what has to be controlled. This means that, not only the violation of Shannon's theorem conditions does not limit the controller performance, but it even helps to reduce the controller complexity. The need for low pass filters to eliminate the ripple from the sampled signal is, in fact, removed. This effect is schematically shown in Fig. 3.4.

We can see that synchronization allows the reconstruction of the average signal value any time the sampling takes place in the middle of the switch on period or in the middle of the switch off period (or both, if double update mode is possible). Instead, if the switching and sampling frequencies are different, low-frequency aliased components will be created in the reconstructed signal. Please note that, even if the sampling and switching frequencies are set equal, there still can be a zero frequency error in the reconstruction of the average sampled signal, in case the sampling instants are not coincident with the beginning and/or the half of the modulation period. This is generally a minor problem, since the current regulator will often be driven by an external loop (see Chapter 6) that, typically including an integral action, will compensate for any steady-state (or very low frequency) error in the current trajectory.

To minimize aliasing effects and reconstruction errors, practically all of the microcontrollers and DSPs designed for power converters control allow the virtually perfect synchronization of the sampling and switching processes. In most cases, the ADC operation is synchronized by the processor hardware with the modulator. Typically, analog to digital conversion of the control variables is started by a signal that also clocks the modulation period beginning and can be re-triggered at a half of the modulation period, if needed.

3.1.3   QUANTIZATION NOISE AND ARITHMETIC NOISE

Quantization of variables and finite arithmetic precision are two among the most critical issues in digital control. Even if a detailed discussion of these issues is far beyond the scope of this book, we feel like it is mandatory to recall at least some basic information about both of them. The interested reader can deepen his or her knowledge of both issues referring to digital control and digital signal processing text books like the very good ones [1, 2] and [3].

As we briefly discussed in Section 3.1.1 quantization takes place any time the amplitude values of a sampled signal are coded using a finite set of symbols. While the original signal's instantaneous amplitude can assume an infinite number of values in a given range, the sampled and coded signal's amplitude can only take one out of a finite number of possible values. The typical implementation of analog to digital conversion in microcontrollers and DSPs associates a *binary* code to the amplitude values of the sampled signal. In the case of the uniform quantizer,

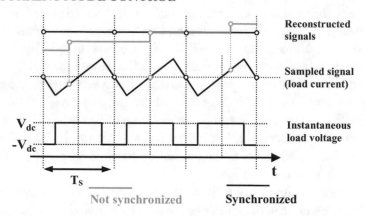

Figure 3.4: Example of synchronized sampling and switching processes. In case the sampling takes place always at the beginning (or in the middle) of the modulation period, the average current value is automatically obtained. If the sampling frequency is lower than the switching one, an aliased, low-frequency component appears on the reconstructed signal.

the rule to associate a binary code N to any given signal sample x is very simple, and can be mathematically expressed as:

$$\begin{cases} \left(N - \dfrac{1}{2}\right) \cdot q < x < \left(N + \dfrac{1}{2}\right) \cdot q & \Rightarrow x_q = N \\[2mm] q = \dfrac{FSR}{2^n} = LSB, \end{cases} \qquad (3.2)$$

where n represents the ADC bit number and, as was previously described, if FSR represents the full scale range, in Volt, of the ADC, then q is the ADC quantization step, equal to one least significant bit (LSB). Please note that (3.2) simply translates the trans-characteristic of the uniform quantizer depicted in Fig. 3.3(a) into a mathematical form. From (3.2) we see that q represents the minimum variation of input signal x that *always* causes the variation of at least one bit in the binary code associated to x_q, the coded signal. Therefore, any variation of signal x smaller than q is not always able to determine some effect on x_q. This simple observation shows us that the quantization process actually implies the loss of some of the information associated to the original signal x. It is a common approach to model this effect as an additive noise, superimposed to the signal. In order to simplify the mathematical characterization of the quantization noise, a stochastic process is associated to it and assumed to be not correlated to signal x, uniform in probability density and with a statistical power equal to:

$$\sigma_q^2 = \frac{LSB^2}{12}. \qquad (3.3)$$

It is then possible to derive a very useful relation that expresses the *maximum* signal to noise ratio (SNR) of an ADC, as a function of its number of bits. This turns out to be given by:

$$SNR = 10 \cdot \log_{10} \left(\frac{12}{8} \cdot 2^{2n} \right) = 6.02 \cdot n + 1.76 \, [\text{dB}]. \tag{3.4}$$

We will not deepen the statistical modeling of the quantization noise any further. Equation (3.4) is a very useful tool to estimate the number of bits one needs, in order to get a desired *SNR* for a given conversion process. For example, if one needs at least a 50 dB *SNR*, (3.4) shows that the number of bits should be higher than 8. Please note that this model does not take into account any other source of noise besides quantization, like, for example, those associated to the signal conditioning circuitry or to the power converter. Consequently, the *actual* signal to noise ratio will always be lower than what is estimated by using (3.4).

There are at least two other major forms of quantization that always take place in the implementation of a digital control algorithm: (i) arithmetic quantization and (ii) output quantization. As far as the former is concerned, we can say that what we call arithmetic quantization is nothing but an effect of the finite precision that characterizes the arithmetic and logic unit used to compute the control algorithm. The finite precision determines the need for truncation (or rounding) of the controller coefficients' binary representations, so as to fit them to the number of bits available to the programmer for variables and constants. In addition, it may determine the need for truncation (or rounding) after multiplications. In general, the effect of coefficient and multiplication result truncation (or rounding) is a distortion of the controller's frequency response, i.e., the shift of the system poles, that can have some impact on the achievable performance. Both truncation and rounding effects can be modeled again as a type of quantization and so as an equivalent noise, of arithmetic nature, added to the signal. Although extremely interesting, predicting the amplification of arithmetic noise within a closed loop control algorithm by pencil and paper calculations is a really tough job. To check the control algorithm operation to this level of detail, the only viable option is its complete, low-level simulation, based on a model that includes the emulation of the adopted controller arithmetic unit.

It should be clear by now that, in case a floating point representation of constants and variables within a control algorithm were employed, none of the above discussed arithmetic quantization effects could be observed. It is important to say, however, that the use of floating point processors in the field of digital control industrial applications is relatively rare. Indeed, only high-end DSPs, designed for high performance real time signal processing, can rely on a floating point arithmetic unit. The cost of such DSP units, however, goes often beyond the maximum affordable for the typical industrial control application. Therefore, at least for the near future, the industrial engineer, designing digital regulators for switching converters, will have to face the problems generated by fixed point arithmetic units. Luckily, the availability of low cost 16- or even 32-bit microcontrollers and DSPs is very large. The occurrence of severe arithmetic quantization problems is therefore hardly ever encountered, being confined to extremely demanding applications or to applications where the use of 8-bit microcontrollers is the only viable option and the emulation of a higher precision arithmetic is out of the question for memory or timing constraints. It is basically for this reason that we will not take arithmetic quantization into account in the following discussion of digital controller implementations. In practice, our results will be determined assuming an infinite precision arithmetic, considering that to be well approximated by modern, 16-bit microcontrollers and DSPs.

Output quantization, instead, is related to the truncation (or rounding) operation inherent in the digital to analog conversion that brings the control algorithm output variable back from the digital to the continuous time domain. In our application case, this function is actually inherent in the digital PWM (DPWM) process. The reduction of the control variable output (in our case the desired duty-cycle) bit number, needed to write it into the PWM duty-cycle register, represents again a quantization noise source. Note that, unless a very high clock to modulation frequency ratio is available (see Section 2.2.2), the effective number of bits that might be used to represent the duty-cycle is always going to be much smaller than the typical variable bit number (16 or 32). Therefore, output quantization is unavoidable. The most unpleasing effect of output quantization may be the occurrence of a peculiar type of instability, specific of digital control loops, that is known as *Limit Cycle Oscillation* (LCO).

3.1.4 LIMIT CYCLE OSCILLATIONS

To open just a brief discussion of LCOs, we would like to show, in the first place, how a limit cycle can be generated in a very simple situation. The case is depicted in Fig. 3.5. We may figure variable d is the duty-cycle of a switching converter, like the one considered in our discussion, whose desired set-point is the particular value we need to apply to bring the converter to the steady state. Variable x may be associated, for example, to the converter average output current. Unfortunately, as we see from Fig. 3.5, the desired set-point for d is not anyone of the possible outputs, because of output quantization.

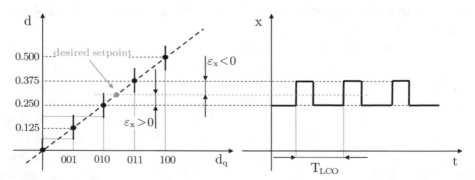

Figure 3.5: Example of limit cycle occurrence. The desired set-point for the control variable d is not anyone of its possible values. Consequently, the system oscillates, with period T_{LCO}. Here we assume that at least one integral action is included in the transfer function from input d to the output x.

As a result, we will in any case apply either a bigger than needed duty-cycle, causing the current increase beyond the steady-state level, or a lower than needed duty-cycle, causing the current decrease below the steady-state value. This happens because the converter output current is, to a first approximation, proportional to the integral of the inverter average output voltage, that is in turn, proportional to the duty-cycle. Commutations between the two states are determined by the current controller, that reacts to the current error build-up by changing the duty-cycle.

This results in a persistent oscillation, i.e., a limit cycle, of the control variables, that is not due to any system instability, but only to the presence of the output quantization. Of course, the amplitude and frequency of the limit cycle are largely dependent on several controller and converter parameters like, for example, controller bandwidth, open loop system time constants and open loop system static gain. Please note that, in cases like the half-bridge converter we are considering here, where the input to output converter transfer function presents a low-pass behavior well approximated by an integral action, this type of limit cycle is practically *unavoidable*.

Within the general digital control theory, limit cycles have been extensively studied, with different degrees of detail and complexity. In power electronics and, more precisely, in the area of DC-DC converter applications, several fundamental papers on quantization resolution and limit cycling have been published, like, for example [4, 5] and others there cited. Without entering too much into this fairly complex topic, we would now like to review the fundamental conditions for the elimination of limit cycles. It is worth clarifying, right from the start, that the conditions reported hereafter are necessary, *but not sufficient*, for the elimination of limit cycle oscillations. Therefore, the actual presence and amplitude of LCOs are usually checked by means of time-domain simulations. This may be a time-consuming investigation, since the presence of LCOs strongly depends on the converter operating point, e.g., on the load current and input voltage levels. In some cases, the system does not show LCOs, unless for a very small set of output current values. In addition, a limit cycle can sometimes be triggered only by some particular transients, having a very particular amplitude. It is therefore not so easy to make sure of the absence of LCOs in *all* the operating conditions of interest.

However, in order to review the fundamental conditions for the existence of a LCO-free steady-state operating point, let us consider the digitally controlled power converter shown in Fig. 3.6(a), where we assume that the dominant quantization effects derive from the ADC and the DPWM, while the rounding effects in the control algorithm are neglected. As a matter of fact, the fixed-point arithmetic and the coefficient round-off may play a relevant role in the accuracy of the controller frequency response definition and in the amplification of quantization noise. Nevertheless, a practical design approach is often based on the assumption of infinite controller arithmetic resolution and on the verification, *a posteriori*, by means of time-domain simulations and experiments.

The *first* condition is to ensure that the variation of one DPWM level, i.e., $1\,LSB$ of the duty-cycle digital representation, here denoted as q_{DPWM}, does not yield a variation of the controlled output variable $x(t)$, in steady-state conditions, greater that the quantization step of $x(t)$,

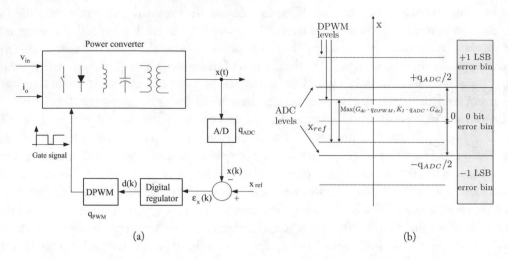

(a)   (b)

Figure 3.6: (a) Digitally-controlled power converter with ADC and DPWM quantization and (b) quantization of state variable $x(t)$ and effects of DPWM quantization.

here denoted as q_{ADC}. Thus, if we define as $G(s)$ the transfer function between the duty-cycle, d, and the controlled variable, $x(t)$, the first necessary condition for the existence of a LCO-free steady state operating point is:

$$q_{DPWM} \, G_{dc} < q_{ADC}, \qquad (3.5)$$

where G_{dc} is the steady-state gain (i.e., $G_{dc} = |G(j0)|$, typically > 1). Condition (3.5) indicates that the DPWM resolution must be greater than the ADC resolution. This guarantees that the DPWM, by itself, has enough resolution to steadily keep the quantized regulation error inside its zero bin. It is worth noting that this reasoning applies to the control of DC quantities, while the analysis, and even the interpretation, of limit cycles in the presence of time-varying references (as in DC-AC converters) may be significantly different.

The *second* necessary condition is the presence of an integral component in the controller. This condition has been formally demonstrated in [5]. However, its motivation can be explained considering that, if only a proportional term (or a proportional-derivative term) is included in the adopted controller, a minimum quantized error on the controlled variable $x(t)$ determines a variation on the controller output that is equal to $K_P \cdot q_{ADC}$. This will allow the control loop to keep a steady operating point only if the following condition is satisfied

$$K_P \cdot q_{ADC} < q_{DPWM}, \qquad (3.6)$$

which is not always the case, because K_P is usually greater than unity. Condition (3.6) simply says that, in order to guarantee the existence of a LCO-free steady-state operating point, the control output must vary by smaller steps than the duty-cycle. Obviously, (3.6) is easier to satisfy when a lower amplification of the minimum quantized error of the input variable can be ensured. This always happens when an *integral component* is included in the control algorithm. In that case, the steady-state operating point is, by construction, inside the zero bin of the controlled variable error. This makes the proportional gain of the controller irrelevant. The integral gain, instead, *induces* a smaller quantization effect on the DPWM, since the minimum variation of the duty-cycle, due to the minimum quantized error on $x(t)$, is now equal to $K_I \cdot q_{ADC}$, with K_I normally much smaller than K_P. As a result, condition

$$K_I \cdot q_{ADC} < q_{DPWM} \tag{3.7}$$

is much more easily met than (3.6). Together with condition (3.5), (3.7) guarantees that it is possible, although *not always* possible, to avoid the occurrence of LCOs. It is immediate to show that the simultaneous verification of both these conditions yields the following one

$$K_I \, G_{dc} < 1, \tag{3.8}$$

that reveals the existence of an upper limit for the integral gain K_I. Summarizing what we have just found, in order to make the elimination of LCOs *theoretically possible* the simultaneous verification of conditions (3.5) and either (3.6) or (3.7) is necessary. To unify the necessary conditions, we can define the DPWM quantization step as the *maximum* between the *physical,* hardware quantization and what we have called the *induced* quantization, determined, e.g., by the integral term, as per (3.7). We can then impose that step to be smaller than q_{ADC}. A schematic representation of these considerations is given in Fig. 3.6(b).

However, it is worth underlining once more that, even if the two conditions above are satisfied, limit cycle oscillations may still be present, essentially because of the effect of the quantizer nonlinearity on the feedback loop. This possible instability may be analysed using *describing function techniques*, including the ADC quantization and possibly the DPWM's one. Thus, the *third* condition for LCO elimination is that the closed-loop system is stable from the describing function's standpoint. Unfortunately, the describing function approach is reasonably accurate only in the case of limit-cycle oscillations that are well approximated by sinusoidal waveforms.

In conclusion, we can say that the analytical prediction of the occurrence of limit cycles, of their amplitude and their frequency is a very complicated problem. In any case, the use of simulation is highly recommended, since the compliance with the above conditions, as we explained, does not guarantee the absence of LCOs. However, it is important to underline that, even if a limit cycle is detected, a proper design of the controller and the signal acquisition path can generally bring its amplitude and frequency to practically acceptable levels.

3.2 BASIC DIGITAL CURRENT CONTROL IMPLEMENTATIONS

In this section we present the basic implementations of the digital current controller for the VSI depicted in Fig. 3.1. We are going to discuss different control algorithms and the related design criteria, with the intention of highlighting the merits and the limitations of each solution. The discussion will refer to an ideal controller implementation, where the above mentioned quantization effects can be considered negligible. Instead, we will focus our attention on the performance allowed by the different solutions and on the impact of the digital controller implementation on the dynamic response of the converter, considering, in particular, figures of merit like the response delay to step changes in the current reference, or the residual tracking error in the presence of sinusoidal reference current signals. Throughout the discussion, we will refer to the converter parameters that we have already taken into account in Sec. 2.3.1, where we presented the analog controller implementation, and that are reported in Table 2.1.

3.2.1   THE PROPORTIONAL INTEGRAL CONTROLLER: OVERVIEW

The first digital controller we want to discuss is the proportional integral, or PI, controller. In the last part of Chapter 2, we have been describing in detail a possible analog implementation of this solution. We now move to a digital implementation observing that, in general, it can be quite convenient to derive a digital controller from an existing analog design. This procedure, that is called *controller discretization*, has the advantage of requiring only a minimal knowledge of digital control theory to be successfully applied. All that is needed is a satisfactory analog controller design and the application of one of the several possible discretization methods to turn the analog controller into a digital one. As we will see in the following, although generally satisfactory, the application of this method implies some loss of precision, as compared to a direct digital design, mainly due to the approximations involved in the discretization process itself and in the equivalent continuous time representation of delays.

Referring to Fig. 3.7, we can see the block diagram of the control loop. As can be seen, it replicates the organization of the block diagram of Fig. 2.10, with the remarkable difference that some of the blocks are now *discrete time* blocks. In particular, we can see how the controller and modulator blocks are now inside the digital domain, the shaded area, that represents a microcontroller or DSP board. The inverter and transducer models are instead exactly equal to those of Fig. 2.10, and, as such, continuous time models. The link between the two time domains is represented by the ideal sampler at the input of the controller and by the digital pulse width modulator, that generates the controller output and, as we have explained, inherently implements the interpolator, or holder, function. All these characteristics imply that we are actually dealing with a *sampled data* dynamic system.

For the reasons we previously explained talking about synchronization, we assume the controller operation is clocked by the DPWM, i.e., a new iteration of the control algorithm is started as soon as a modulation period begins. We also assume, for simplicity, that the single update mode of operation is adopted, so that, during each modulation period, a single new value of the controller output is computed. The computation is based on the current sample, acquired at the start of the period and indicated by $I_O^S(k)$. Since the controller operation proceeds at time steps that are multiple of T_S, the modulation and sampling period, in all the controller signals we simply denote with k the instant $k \cdot T_S$ from the origin of time. Accordingly, we say that, at the k-th modulation period, the output of the controller, i.e., the digital modulating signal, is $m(k)$. Please note that, although we keep identifying the output of the controller by m, as in the analog case, this must no longer be considered an analog signal, but rather a sequence of binary codes, i.e., a quantized discrete time signal. Of course, the same holds for each of the other controller internal signals, like I_{OREF} and I_O^S.

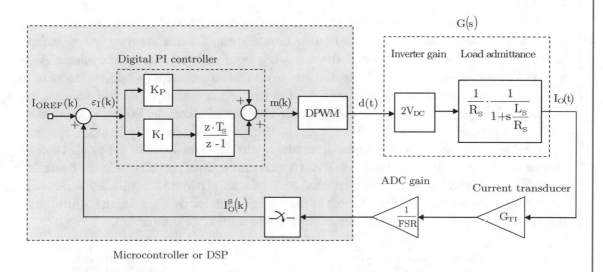

Figure 3.7: Block diagram of the digital current control loop with PI regulator.

It is worth noting that, in order to make Fig. 3.7 more realistic, we are going to modify the static gains of the modulator and of the feedback path with respect to the analog design example of Fig. 2.10. Indeed, in a digital implementation, the modulator static gain is represented by the numerical scale factor that turns the binary code $m(k)$ in the corresponding duty-cycle $d(t)$. In general, this depends on the way variables are normalized in the control algorithm. It is possible to verify that, as soon as the normalization of variables is such that $m(k)$ is coded as a fractional binary number, i.e., the maximum binary value of m is made equivalent to unity, then the modulator static gain is also unity, i.e., $m(k)$ directly represents the duty-cycle, without further scale factors. The fractional normalization hypothesis also explains the presence of the ADC gain at the input of the digital controller, meaning that a full scale input value of the ADC is normalized to unity as well. Under these assumptions and without loss of generality, we will assume the DPWM static gain to be equal to unity. If a different normalization criterion is adopted, the modulator static gain will have to be adjusted accordingly.

3.2.2   SIMPLIFIED DYNAMIC MODEL OF DELAYS

As briefly outlined above, the application of discretization techniques requires the designer to determine an equivalent continuous time model of his or her sampled data system, to use that in the design of a continuous time controller stabilizing the feedback loop and, finally, to turn the continuous time controller into an equivalent discrete time one. Therefore, first of all, we need to discuss the derivation of an equivalent, continuous time model for the system represented in Fig. 3.7.

The typical text book approach [2, 3] to sampled data dynamic systems control normally requires to properly model, in the continuous time domain, the discrete time system included between the ideal sampler located at the controller input and the output interpolator. As we have explained in Section 3.1.1, the typical way to do this is considering a suitable model of the interpolator, e.g., some kind of holder, and, after that, finding an equivalent continuous time representation for the cascade connection of the ideal sampler and the holder, that is called a *sample and hold*. Please note that this method, schematically illustrated by Fig. 3.8, is actually what we have already used in Chapter 2, modeling the different types of DPWM. Once the sample and hold is modeled, the designer can operate the controller synthesis in the continuous time domain, assuming that, once converted back into a discrete time equivalent and inserted between the sampler and the interpolator in the original sampled data system, the controller will maintain the closed loop properties determined by the continuous time design.

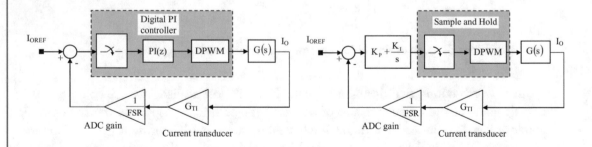

Figure 3.8: Procedure to define the continuous time equivalent of the digital current control loop.

This is what we have to do with the sampled data system of Fig. 3.7, with a significant difference. The difference lies in the fact that, in this case, the function of the interpolator is inherent to the DPWM, because that is the block where the conversion from the digital to the analog domain takes place. This means that once the holder effect is properly modeled in the DPWM the conversion of the sampled data system into an equivalent, continuous time one will be complete. This may seem a minor detail, but in this difference lies the key for the correct interpretation of the system in Fig. 3.7 as a sampled data system. In Chapter 2, we have described several continuous time equivalent models for the DPWM. Considering, for example, model (2.7), after minor rearrangements and assuming, as we explained above, $c_{PK} = 1$, we get the following expression:

$$DPWM(s) = \frac{1}{2}\left(e^{-s(1-D)\frac{T_S}{2}} + e^{-s(1+D)\frac{T_S}{2}}\right) = e^{-s\frac{T_S}{2}}\cos\left(\omega\frac{T_S}{2}D\right) \cong e^{-s\frac{T_S}{2}}, \qquad (3.9)$$

that, as can be seen, shows the equivalence of the considered DPWM to a half modulation period delay, cascaded to a frequency-dependent gain. Considering the typical current controller bandwidth to be limited well below the modulation frequency, $1/T_S$, the gain term can actually be approximated by unity, independently from the duty-cycle D, so that the last part of (3.9) holds. In the above assumptions, (3.9) shows that we can quite accurately model the DPWM as a pure, half modulation period delay. Please note that this exactly coincides with the continuous time model of the zero-order hold usually adopted in a sampled data controller design. Of course, if a different DPWM model were considered, the result (3.9) would represent a coarser approximation, but could still be used as a simplified representation of the holder delay effect. Considering now a first-order Padé approximation of (3.9), a rational, continuous time transfer function can be obtained, whose expression is the following:

$$e^{-s\frac{T_S}{2}} \cong \frac{1 - s\dfrac{T_S}{4}}{1 + s\dfrac{T_S}{4}}, \qquad (3.10)$$

where T_S is, of course, the sampling period. The usefulness of (3.10) is that a rational transfer function is clearly easier to deal with than the exponential function. We have actually already met (3.10) in Chapter 2, Fig. 2.10, where it was used, basically under the same assumptions, to approximately model the DPWM delay in an analog regulator design example.

We are now ready to consider the continuous time equivalent of our sampled data system. This is shown in Fig. 3.9. As can be seen, we have obtained exactly the same model of Fig. 2.10, with the only difference that the static gain of the modulator is now considered equal to one and that there is an additional gain in the feedback path. To simplify the following developments of this result, we assume $FSR = c_{PK}$, so that that the open loop static gain of Fig. 3.9 and that of Fig. 2.10 are identical. Of course, in general, the two loop gains will have a different DC value, which will require some straightforward adjustment of the controller parameters. Under our assumption instead, the analog PI controller we have designed in Chapter 2, represents a satisfactory stabilizing controller also for the loop of Fig. 3.9.

Therefore, we are now ready to take the last step towards the design of the digital PI current controller. All we have to do is apply a suitable discretization method to the analog controller we already possess. The way this can be done is the subject of the next section.

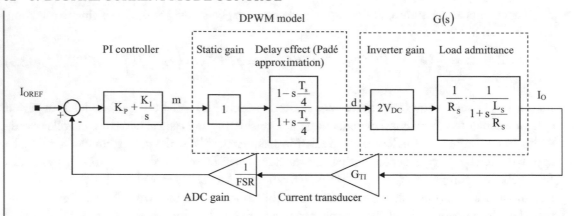

Figure 3.9: Block diagram of the continuous time equivalent of the digital current control loop.

3.2.3   THE PROPORTIONAL INTEGRAL CONTROLLER: DISCRETIZATION STRATEGIES

According to digital control theory, the application of any discretization method always implies a loss of performance with respect to a purely analog control implementation. This is true also for our case. Since the delay effect of the analog PWM is negligible with respect to that of a digital modulator, if an analog current controller were designed for the system of Fig. 3.7, the control loop bandwidth could indeed be higher.

In Chapter 2 we have chosen to design the analog PI controller considering a digital PWM modulator and modeling its delay exactly as in Fig. 3.9. That choice, together with the "educated" choice of the ADC FSR value that was done in the previous paragraph, allowed us to find a controller that, although not ideal for the analog implementation, is now ready for discretization without further adjustments. From a textbook's standpoint, this offers two advantages: to keep the presentation more compact and to allow, in the end, the comparison of two virtually identical controllers, analog and digital, and thus putting into evidence the impact of discretization on the final performance. However, note that, in the general case, the analog design would have to be started from scratch, based on the equivalent model of Fig. 3.8.

There are actually several possible discretization strategies, some based on the invariance of the dynamic response to particular signals (steps, ramps, etc.), and others based on numerical integration methods. The latter are the ones we are going to consider now. The basic concept behind them is very simple: we want to replace the continuous time computation of integrals with some form of numerical approximation. The two basic methods that can be applied at the purpose are known as Euler integration and trapezoidal integration method. The principle is illustrated in Fig. 3.10.

As can be seen, the area under the curve is approximated as the sum of rectangular or trapezoidal areas. The Euler integration method can actually be implemented in two ways, known as *forward* and *backward* Euler integration, the meaning being obvious from Fig. 3.10(a). Writing the rule to calculate the area as a recursive function of the signal samples, applying $\mathcal{Z}$-transform to this area function, and imposing the equivalence with the Laplace transform integral operator, gives a direct transformation from the Laplace transform independent variable s to the $\mathcal{Z}$-transform independent variable z.

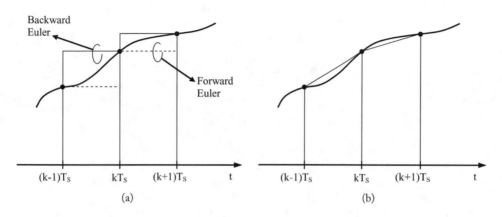

Figure 3.10: (a) Euler integration method (forward and backward) and (b) trapezoidal integration method.

Table 3.1 shows the transformations that are obtained for the two discretization methods, where the two possible versions of the Euler integration method are considered. These are called *Z-forms*. The practical meaning of each Z-form is the following: the substitution of the s variable in the controller transfer function with the indicated function of the z variable, determines the transformation of the continuous time controller into an *equivalent* discrete time one, the equivalence being in the sense of the integral approximation explained above.

Table 3.1: Discretization methods

METHOD	Z-FORM	3% DISTORTION LIMIT
Backward Euler	$s = \frac{z-1}{zT_s}$	$\frac{f_s}{f} > 20$
Forward Euler	$s = \frac{z-1}{T_s}$	$\frac{f_s}{f} > 20$
Trapezoidal (Tustin)	$s = \frac{2}{T_s}\frac{z-1}{z+1}$	$\frac{f_s}{f} > 10$

Aside 3. Discretization of the PI Current Controller

In Aside 2, we have determined the proportional and integral gains of an analog PI current controller. These were $K_P = 2.01$ and $K_I = 2.16 \times 10^3$ (rad s^{-1}). The corresponding controller transfer function is given by

$$PI(s) = K_I \frac{1 + s \cdot \dfrac{K_P}{K_I}}{s}.$$ (A3.1)

We now proceed to the controller discretization, considering, at first, the Euler integration method in the *backward* version. Substituting the s variable with the expression indicated in the first row of Table 3.1, we find

$$PI(z) = K_I \frac{1 + \dfrac{z-1}{z \cdot T_S} \cdot \dfrac{K_P}{K_I}}{\dfrac{z-1}{z \cdot T_S}} = \frac{(K_P + K_I \cdot T_S) \cdot z - K_P}{z - 1} = K_P + K_I \cdot T_S \cdot \frac{z}{z-1}.$$

(A3.2)

As can be seen, we have obtained a new rational transfer function that can be simplified to give the discrete time implementation of the PI controller. The block diagram corresponding to the last expression in (A3.2) is shown in Fig. A3.1, which represents the parallel realization of the discrete time regulator, followed by a possible, very simple, model of the calculation delay.

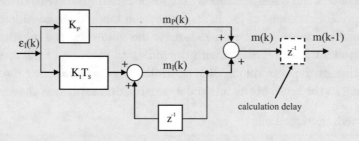

Figure A3.1: Block diagram representation of the digital PI controller.

Recalling the basic $\mathcal{Z}$-transform properties, we can immediately write down the control algorithm that may be used to implement the PI regulator in our microcontroller or DSP unit. This is as follows:

$$\begin{cases} m_I(k) = K_I \cdot T_S \cdot \varepsilon_I(k) + m_I(k-1) \\ m(k) = m_P(k) + m_I(k) = K_P \cdot \varepsilon_I(k) + m_I(k), \end{cases} \tag{A3.3}$$

where $\varepsilon_I(k)$ represents the current error at instant kT_S. Please note that Fig. A3.1 actually represents a more detailed description of the digital PI controller depicted also in Fig. 3.7.

Similarly, we can apply the trapezoidal integration based Z-form, also known as Tustin transform. Following the same procedure above, it is easy to derive the control algorithm that translates the discretized PI controller. We find

$$\begin{cases} m_I(k) = K_I \cdot T_S \cdot \dfrac{\varepsilon_I(k) + \varepsilon_I(k-1)}{2} + m_I(k-1) \\ m(k) = m_P(k) + m_I(k) = K_P \cdot \varepsilon_I(k) + m_I(k). \end{cases} \tag{A3.4}$$

As can be seen, the structure of (A3.4) is similar to that of (A3.3); the only difference being determined by the computation of the integral part that is not based on a single current error value, but rather on the moving average of the two most recent current error samples.

This difference is responsible for the lower frequency response distortion of the Tustin transform. It is worth noting that the proportional and integral gains for the two different versions of the discretized PI controller are exactly the same. As can be seen, in both cases we find that the proportional gain for the digital controller is exactly equal to that of the analog controller, while the digital integral gain can be obtained simply by multiplying the continuous time integral gain and the sampling period. Please note that also the application of pre-warping does not change much the values of the controller gains; especially when a relatively high ratio between the sampling frequency and the desired crossover frequency is possible. This is also confirmed by the Bode plots, shown in Fig. A3.2, that refer to each of the different PI controllers we have considered so far, i.e., the original continuous time one and of each of the three discretized versions (Euler, Tustin, and pre-warped).

As can be seen, with our design parameters and sampling frequency, the plots are practically indistinguishable.

In summary, we have seen that, given a suitably designed analog PI regulator, the application of any of the considered discretization strategies simply requires the computation of the digital PI gains, as in the following:

$$\begin{cases} K_{I\_dig} = K_I \cdot T_S \\ K_{P\_dig} = K_P, \end{cases} \tag{A3.5}$$

and the implementation of the proper control algorithm (A3.3) or (A3.4).

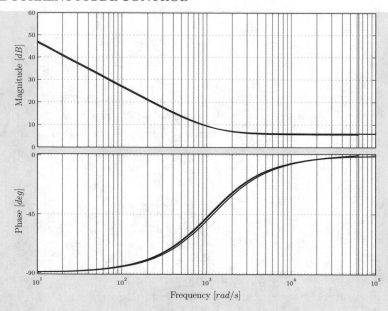

Figure A3.2: Bode plots of the different PI realizations. The vertical line indicates the Nyquist frequency.

The last issue we need to discuss is the role of the calculation delay model that appears in Fig. A3.1 (dotted z^{-1} block). If the unit delay block is added to the controller block diagram, it becomes possible to evaluate the effect of the calculation delay on the control performance and the closed loop system stability. This can be done using any kind of system modeling and simulation software. Of course, the duration of the calculation delay is, in this case, supposed to be equal to one sampling period, as a worst-case approximation. More importantly, the design of the original analog PI controller was performed neglecting the calculation delay, so it is likely that its inclusion in the digital controller model, at the time of verification, will significantly affect the dynamic performance. To compensate that, the analog design should be corrected considering an equivalent control loop-delay equal to $(3/2)T_S$ in (3.10).

Since the numerical integration methods imply a certain degree of approximation, if we compare the frequency response of the controller before and after discretization, some degree of distortion, also known as frequency warping effect, can always be observed. Table 3.1 also shows the condition that has to be satisfied to make the distortion lower than 3% at a given frequency f. The condition is expressed as a limit for the ratio between the sampling frequency $f_S = 1/T_S$ and the frequency of interest, f. As can be seen, the trapezoidal integration method, that generates the so-called Tustin Z-form, is more precise than the Euler method, and, as such, guarantees a smaller distortion at each frequency or, equivalently, a higher 3% distortion limit, that is as high as one tenth of the sampling frequency. Ideally, it is also possible to pre-warp the controller transfer function so as to compensate the frequency distortion induced by the discretization method and get an exact phase and amplitude match of the continuous time and discrete time controllers at *one* given frequency, that is normally the desired crossover frequency.

However, this method implies some more involved calculations and is therefore easily applicable only if we can use some mathematical software implementing the discretization techniques. As we show in Aside 3, in the typical application case, the difference in the controller frequency response we can get is usually small, so that the application of discretization methods more complex that the Euler one is very seldom motivated, at least for the PI controller.

To conclude the discussion of discretization techniques, we now present the results of the simulation and experimental test of our VSI with the digital controller obtained following the procedure outlined in the Aside 3 and implementing the algorithm described by (A3.3). They are shown in Fig. 3.11. It is important to mention that, in order to achieve a negligible calculation delay, in agreement with the above discussed model, a FPGA controller implementation, with 33 ns calculation delay, has been used in the experiments. This makes the trace of Fig. 3.11(d) comparable with the simulation result of Fig. 3.11(c), that was computed in the same set-up conditions, and explains the really good match between the simulated and measured dynamic responses.

It is interesting to compare these results with those reported in Fig. 2.11. As can be seen, there is really little difference in the performance achieved by the discretized controller. Indeed, it is possible to notice just a small increase in the phase shift between the output current and its reference, a consequence of the slightly lower bandwidth achieved by the digital controller, as well as an equally small reduction of the response damping, due to a somewhat lower phase margin (both effects of the approximations implicit in the design procedure and of the resulting frequency warping).

3.2.4   EFFECTS OF THE COMPUTATION DELAY

In the above discussion, we showed how the delay effect associated to the DPWM operation can be taken care of. An additional complication we have to deal with is represented by the fact that the block diagram of Fig. 3.7 actually hides a second, independent source of delay: this is the control algorithm computation delay, i.e., the time required by the processor to compute a new m value, given the input variable sample. Although digital signal processors and microcontrollers are getting faster and faster, in practice the computation time of a digital current controller always represents a significant fraction of the modulation period, ranging typically from 10%–40% of it. A direct consequence of this hardware limitation is that, in general, we cannot compute the input to the modulator during the same modulation period when it has to be applied. In other words, the modulator input, in any given modulation period, must have been computed during the

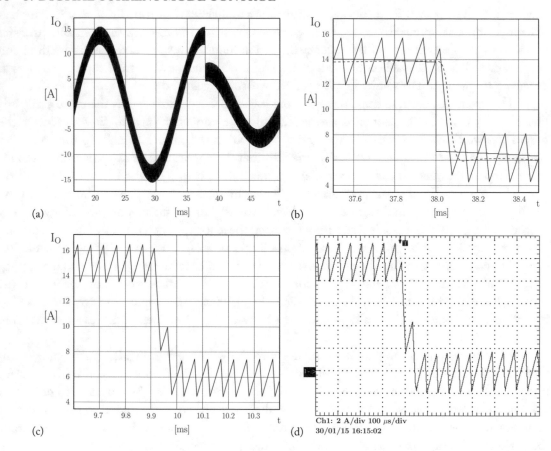

Figure 3.11: Simulation of the VSI with the controller designed according to the procedure reported in the Aside 3. The depicted variable is the VSI output current I_O. (a) Controller response to a step reference amplitude change. (b) Detail of previous figure. It is possible to see that no calculation delay effect has been included in the simulation. (c) Simulation of the experimental set-up step response. (d) Experimental verification of the PI step response with negligible calculation delay.

previous control algorithm iteration. Dynamically, this means that the control algorithm actually determines an additional one modulation period delay.

One could consider this analysis to be somewhat pessimistic, because powerful microcontrollers and DSPs are available today, that allow the computation of a PID routine in much less than a microsecond. However, it is important to keep in mind that, in industrial applications, the cost factor is fundamental: cost optimization normally requires the use of the minimum hardware that can fulfill a given task. The availability of hardware resources in excess, with respect to what is strictly needed, simply identifies a poor system design, where little attention has been paid to the cost factor. Therefore, the digital control designer will struggle to fit his or her control routine to a minimum complexity microcontroller much more often than he or she will experience the opposite situation, where a high speed DSP will be available just for the implementation of a digital PI or PID controller.

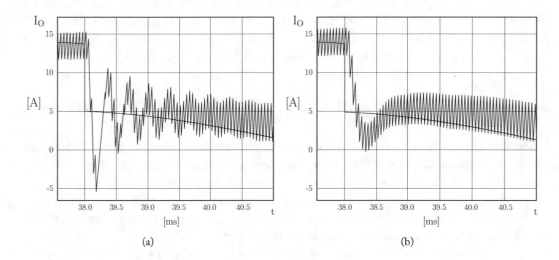

(a)

(b)

Figure 3.12: Simulation of the VSI with the digital PI controller including the calculation delay. (a) Detail of the controller response to a step reference amplitude change without re-design: undershoot and oscillating response. (b) Detail of controller response with re-design: reduced undershoot, reduced speed of response, increase of phase lag.

The conventional approach to tackle the problem consists in assuming a whole control period is dedicated to computations, as shown in the Aside 3, Fig. A3.1. In this case, in order to get from the digital controller a satisfactory performance, the calculation delay effect has to be included from the beginning in the analog controller design. Practically, this can be done increasing the delay effect represented by the Padé approximation of Fig. 2.10 and Fig. 3.9 by T_S. After that, the procedure described in the Aside 3 for the controller synthesis through discretization can be re-applied. It is important to underline once more that, if the analog controller is not re-designed and a significant calculation delay is associated to the implemented algorithm, the achieved performance can be much less than satisfactory. An example of this situation is shown in Fig. 3.12(a), where a calculation delay equal to one modulation period is considered. Note how the step response tends to be under-damped. In the other case instead, as is shown in Fig. 3.12(b), the dynamic response of the re-designed controller is smoother, but a significant reduction of its speed can be observed. Please note that the result has been obtained by reducing the crossover frequency to $f_S/15$, while keeping the same phase margin of the original design. The previous example shows that, when the maximum performance is required, this conventional approach may be excessively conservative. Penalizing the controller bandwidth to cope with the computation delay, the synthesis procedure will unavoidably lead to a worse performance, with respect to conventional analog controllers. This is the reason why, in some cases, a different modeling of the digital controller can be considered, that takes into account the exact duration of the computation delay and so, by using modified $\mathcal{Z}$-transform, exactly models the duty-cycle update instant within the modulation period. In this way, the penalization of the digital controller with respect to the analog one can be minimized and a significant performance improvement, with respect to the case of Fig. 3.12(b), can be achieved. This will be the subject of Section 3.2.6.

3.2.5   DERIVATION OF A DISCRETE TIME DOMAIN CONVERTER DYNAMIC MODEL

What we have described so far is a very simple digital controller design approach. It is based on the transformation of the sampled data system into a continuous time equivalent, that is used to design the regulator with the well known continuous time design techniques. The symmetrical approach is also possible. In this case, the sampled data system is transformed into a discrete time equivalent, that can be used to design the controller directly in the discrete time domain. We are now going to present a short review of this methodology.

Discrete-time models for power electronic circuits have been widely discussed in the past (see, for example, [6, 7] and [8]). The detailed and precise discrete-time converter model is generally based on the integration of the linear and time-invariant state space equations, associated to each switch configuration (i.e., turn-on and turn-off). Then, the state variable time evolutions, obtained separately for each topological or switch state, are linked to one another exploiting the continuity of the state variable, i.e., imposing the final state of one configuration to be the initial state of the next. This approach, that requires the use of exponential matrices, leads to a general discrete-time state-space model and precisely represents the system dynamic behavior in the discrete-time domain. Therefore, in principle, it represents a very good modeling approach for digitally controlled power electronic circuits. Nevertheless, it is not very commonly used, mainly for two reasons: (i) the obtained discrete time model depends on the particular type of modulator adopted, as the sequence of state variable integrations, one for each topological state, depends on the modulator mode of operation (leading edge, trailing edge, etc.); and (ii) the exponential matrix computation is relatively complex and, therefore, not always practical for the design of power electronic circuit controllers.

A more direct, equivalent, approach to discrete time converter modeling is described in Fig. 3.13 and Fig. 3.14(a), where the PWM modulator is represented using the frequency domain model, $PWM(s)$, derived in the previous chapter, $G(s)$, the converter transfer function, is obtained from the continuous-time converter small-signal model, and $x^s(t)$ is the sampled output variable, that has to be controlled by the digital algorithm. To account for the time required by the AD conversion and by the control algorithm computation in the DSP (or microcontroller), a time delay T_d is cascaded to the controller transfer function Reg(z). More explicitly, in a uniformly sampled PWM, time T_d represents the delay between the output variable sampling and the duty-cycle update instants. When this is equal to one modulation period, a simple z^{-1} block could replace it in the control loop.

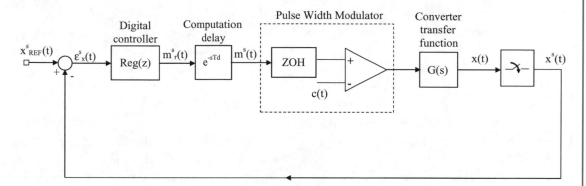

Figure 3.13: Model of the control loop for digitally controlled converters.

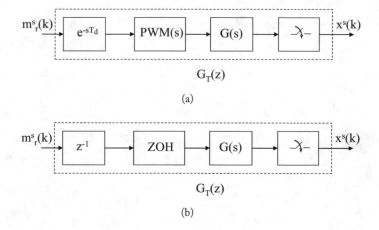

Figure 3.14: Equivalent dynamic model of computation delay, the PWM transfer function, the converter and the sampler: (a) general form and (b) simplified representation, where the PWM is approximated as a Zero-Order-Hold (ZOH) and the control delay is equal one modulation period.

Aside 4. PI Current Controller with Integral Anti-wind-up

In Aside 3, we have completed the design of a digital PI current controller. This Aside is dedicated to a typical implementation issue, i.e., the control of the integral part wind-up. This phenomenon can take place any time the PI controller input signal, i.e., the regulation error, is different from zero for relatively long time intervals. This typically happens in the presence of large reference amplitude variations or other transients, causing inverter saturation. The problem is determined by the fact that, if we do not take any countermeasure, the integral part of the controller will be accumulating the integral of the error for the entire transient duration. Therefore, when the new set-point is reached, the integral controller will be very far from the steady state and a transient will be generated on the controlled variable, which typically has the form of an overshoot. It is fundamental to underline that this overshoot is not related to the small-signal stability of the system. Even if the phase margin is high enough, the transient will always be generated, as it is just due to the way the integral controller reacts to converter saturation. An example of this problem is shown in Fig. A4.1(a).

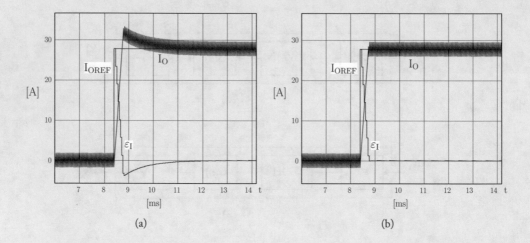

Figure A4.1: Dynamic behavior of the PI controller during saturation: (a) no anti-wind-up and (b) anti-wind-up.

The solution to this problem is based on the dynamic limitation of the integral controller output during transients. Transients can be detected monitoring the output of the controller proportional part: in a basic implementation, any time this is higher than a given limit, the output of the integral part of the controller can be set to zero. Integration is resumed only when the regulated variable is again close to its set-point, i.e., when the output of the proportional part gets below the specified limit. More sophisticated implementations of this concept are also possible, where the limitation of the integral part is done gradually, for example keeping the sum of the proportional and integral outputs in any case lower than or equal to a predefined limit. In this case, shown in Fig. A4.2, at each control iteration, a new limit for the integral part is computed and, if needed, the integral output is clamped.

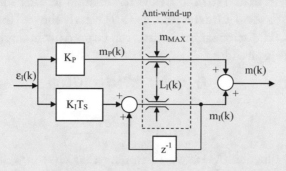

Figure A4.2: Block diagram representation of the digital PI controller with anti-wind-up action.

This implementation, of course, requires a slightly higher computational effort, which amounts to the determination of the following quantity, where m_{MAX} is the controller output limit:

$$|L_I(k)| = m_{MAX} - |K_P \varepsilon_I(k)|. \qquad (A4.1)$$

However, the result can be quite effective, as shown in Fig. A4.1(b). Please note that similar provisions can be as well adopted to limit the state variables of any other type of digital regulator.

Before continuing with this discussion, we ought to clarify two key points, fundamental to establish a correct relationship between our modeling approach and standard digital control theory. First of all, the Zero Order Hold (ZOH) function that, when cascaded to an ideal sampler, models the conversion from sampled time variables into continuous time variables, is, in our case, internal to the PWM model, and, therefore, does not appear right after the sampler. As a consequence, recalling that an ideal sampler has, by itself, a gain equal to $1/T_S$ [1], if we want to correctly represent the transfer function between the sampled time input variable and the continuous time output variable of the modulator, a gain equal to T_S has to be added to the modulator transfer function $PWM(s)$. Clarified this, the discrete time transfer function $G_T(z)$, that exactly represents the discrete-time state variable dynamic equations is given by:

$$G_T(z) = \mathcal{Z}\left[e^{-sT_d} T_s PWM(s) G(s)\right]. \qquad (3.11)$$

This $\mathcal{Z}$-domain approach is very powerful: indeed it is capable of correctly quantifying the difference in the converter dynamics determined by the different uniformly sampled modulator implementations (trailing edge, leading edge, triangular carrier modulation, etc.), as it takes into account the exact duty-cycle update instant. Nevertheless, there are two strong motivations to simplify the discretization process and the evaluation of (3.11) in the case of triangular carrier modulation: (i) the control delay is usually equal to one modulation period, and a simple z^{-1} block can be used to represent it; and (ii) the PWM modulation transfer function $T_S \cdot PWM(s)$ looks very much like that of a Zero-Order-Hold, as (3.9) clearly shows. Therefore, an approximated, but more intuitive, ZOH discretization method can be used to obtain the open loop discrete time transfer function. This is given by

$$G_T(z) = z^{-1} \mathcal{Z}[H(s)\ G(s)], \tag{3.12}$$

where

$$H(s) = \frac{1 - e^{-sT_S}}{s}$$

is the ZOH transfer function. Moreover, assuming that $G(s)$ is well approximated by a pure integrator, as is the case of our current control example with $R_S = 0$, i.e.,

$$G(s) = \frac{2\,V_{DC}}{s\,L_S}$$

and assuming $T_d = T_S$, then there is no difference between (3.11) and (3.12). In fact, re-writing (3.12) we find:

$$G_T(z) = z^{-1} \mathcal{Z}\left[\frac{1 - e^{-sT_S}}{s} \frac{2\,V_{DC}}{s\,L_S}\right]$$

$$= \frac{2\,V_{DC}}{L_S} z^{-1}(1 - z^{-1})\mathcal{Z}\left[\frac{1}{s^2}\right] = \frac{2\,V_{DC}\,T_S}{L_S} \frac{1}{z(z-1)} \tag{3.13}$$

while, re-writing (3.11), we get:

$$G_T(z) = \mathcal{Z}\left[e^{-sT_S}\ T_S\ PWM(s) \frac{2\,V_{DC}}{s\,L_S}\right]$$

$$= \frac{2\,V_{DC}\,T_S}{L_S} z^{-1}\mathcal{Z}\left[\frac{PWM(s)}{s}\right] = \frac{2\,V_{DC}\,T_S}{L_S} \frac{1}{z(z-1)}, \tag{3.14}$$

where $PWM(s)$ is the transfer function given by (2.7), with $c_{PK} = 1$. The equivalence between the two approaches is easily justified if we consider that the output current variation only depends on the integral of the inverter voltage, i.e., in other words, only on the average voltage value generated by the PWM, and not on the particular allocation of the PWM pulse within the modulation period.

Following the same reasoning, the extension of the z-domain small-signal model derivation to the case of the multi-sampled system, described in Section 2.2.4, is straightforward:

$$G_T(z) = \mathcal{Z}\left[e^{-sT_d} \frac{T_S}{N}\ PWM(s)\ G(s)\right], \tag{3.15}$$

where the $\mathcal{Z}$-transform is taken with a sampling period equal to T_S/N. Model (3.14) can be used for the direct discrete time design of the current controller, simply deciding the closed loop poles allocation.

3.2.6   MINIMIZATION OF THE COMPUTATION DELAY

As previously described, the computational delay between the sampling instant and the duty-cycle update instant is one of the most important factors that limit the dynamic performance of digitally-controlled power converters. We have seen before that, in order to avoid anti-aliasing filters, a common practice is to sample the inductor current in the middle of either the turn-on or the turn-off intervals, thus ensuring its average value is automatically acquired. However, this provision usually introduces a delay in the control loop, strongly limiting the achievable bandwidth. The control delay can be reduced by half in the double-update mode, where the input variables are sampled in the middle of both the turn-on and the turn-off interval and the duty-cycle is updated *twice* in each switching period.

However, the increase of computational power of DSPs, microcontrollers, and FPGAs, which are now able to complete the control algorithm computation in smaller and smaller fractions of the switching period, makes possible the further reduction of the control delay. This can be obtained shifting the current sampling instant towards the duty-cycle update instant, leaving just enough time for the ADC to generate the new input sample and to the processor for the control algorithm calculation.

The situation under investigation is depicted in Fig. 3.15, where T_d is, once again, the time required by AD conversion and calculations. Time T_C is instead available for other non-critical functions or external control loops. As can be seen, since $T_d << T_S$, being T_S the modulation period, the sampling of the state variable $x(t)$, i.e., in our case of the inductor current $I_O(t)$, is delayed with respect to conventional controller organizations and shifted towards the PWM update instant, as shown in Fig. 3.15. From the controller's standpoint, this implies a reduction of the feedback loop delay.

In order to quantify the effectiveness of this reduction, an accurate discrete-time model is needed. To this purpose, we can consider the block diagram of Fig. 3.14(a), and replace the PWM block with a Zero-Order-Hold (ZOH), which, as we have seen, represents a very good approximation, especially in the case of triangular carrier waveform. Now, if the control delay T_d is a sub-multiple of the sampling period T_S, the continuous system is easily convertible into a discrete-time model using conventional $\mathcal{Z}$-transform and considering T_d as the sampling period. In our case, the delay T_d is a generic fraction of sampling period T_S and therefore, modified $\mathcal{Z}$-transform has to be used to correctly model the system. The basics of modified $\mathcal{Z}$-transform are briefly recalled hereafter. Let us define:

$$p = 1 - \frac{T_d}{T_S}, \tag{3.16}$$

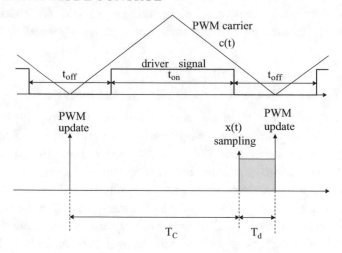

Figure 3.15: Sampling of variable $x(t)$ shifted towards the PWM update.

where $0 \leq p \leq 1$. If $g(t)$ is the impulse response of $G(s)$, we denote, as we did before, the $\mathcal{Z}$-transform of the ideally sampled version of $g(t)$ (i.e., $\mathcal{Z}\left[\mathcal{L}^{-1}\left[G(s)\right](k\, T_S)\right]$ simply as $\mathcal{Z}\left[G(s)\right]$, with $\mathcal{L}^{-1}$ being inverse of Laplace transform. Consequently, the discrete-time model of the continuous system of Fig. 3.7, can be expressed as:

$$\mathcal{Z}\left[\underbrace{H(s)\, G(s)\, e^{-s\, p\, T_S}}_{G_1(s)}\right] = \sum_{k=0}^{\infty} z^{-k} g_1(k T_S - T_d) = \mathcal{Z}_m\left[G_1(s)\right] = G_1(z, p), \qquad (3.17)$$

where $g_1(t)$ is the impulse response of $G_1(s)$, and $G_1(z, p)$ (or $\mathcal{Z}_m[G_1(s)]$) is the modified $\mathcal{Z}$-transform of $G_1(s)$. In the particular case of the Zero Order Hold, $H(s) = (1 - e^{-sT_S})/s$ and (3.17) becomes:

$$G_T(z, p) = \mathcal{Z}\left[\underbrace{\frac{1 - e^{-sT_S}}{s}}_{H(s)} G(s) e^{-s\, p\, T_S}\right] = \frac{z-1}{z}\mathcal{Z}\left[\frac{G(s)}{s} e^{-s\, p\, T_S}\right] = \frac{z-1}{z}\mathcal{Z}_m\left[\frac{G(s)}{s}\right].$$

$$(3.18)$$

The modified $\mathcal{Z}$-Transform maintains the properties of the conventional $\mathcal{Z}$-transform, since it is simply defined as the $\mathcal{Z}$-transform of a delayed signal, see (3.17). The results of the modified $\mathcal{Z}$-transform application to particular cases of interest are usually available in look-up tables [9].

In our example case, the discrete-time transfer function between the modulating signal $M(z)$, input of the DPWM, and the *delayed* inductor current $I_O(z)$ can be written as:

$$\frac{I_O(z)}{M(z)} = \frac{2\,V_{DC}\,T_S}{L_S} \cdot \frac{z\,p - (p-1)}{z(z-1)}. \tag{3.19}$$

It may be interesting to observe that in the usual case, where $p = 0$, i.e., the sampling and computation delay amounts to one full modulation period, (3.19) reduces to

$$\frac{I_O(z)}{M(z)} = \frac{2\,V_{DC}T_S}{L_S} \cdot \frac{1}{z(z-1)}, \tag{3.20}$$

which, as can be verified, is equal to (3.13) and (3.14). In order to quantify the advantages of exactly modeling the delay, i.e., of considering $p > 0$, let us take in to account, as a benchmark parameter, the achievable current loop bandwidth. We assume, for simplicity, that the current regulator is purely proportional and that the phase margin is kept constant, for example to $+50°$. To this purpose, we look for the frequency where (3.19) shows a $-130°$ phase rotation, and we define that as the achievable current loop bandwidth (BW_i). The results are reported in Table 3.2. Please note how, simply by shifting the sampling instant towards the duty-cycle update instant, a significant improvement in the achievable current loop bandwidth can be obtained. It also possible to note that only with $p = 0$ (sampling in the middle of turn-off time) or $p = 0.5$ (sampling in the middle of turn-on time), the sampled current is also the average inductor current, while, for other values of p, some kind of algorithm is needed for the compensation of the current ripple, possibly accounting for dead-time effects as well. For this reason, the application of the concept here described to current control is fairly complicated, while it can be much more convenient for the control of other system variables, where the switching ripple is smaller. This can be the case, for example, of output voltage control in an Uninterruptible Power Supply, a particular application of the VSI we will discuss in Chapter 6.

Table 3.2: Achievable current loop bandwidth BW_i, vs. p[a]

p	0	0.5	0.8
BW_i	$f_s/13.4$	$f_s/9$	$f_s/6.2$

[a] Phase margin is 50° and the current regulator is purely proportional.

3.2.7   THE PREDICTIVE CONTROLLER

We now move to a totally different control approach, describing the predictive, or *dead-beat*, current control implementation [10, 11]. The basic organization of this controller is shown in

Fig. 3.16: it closely resembles the one shown in Fig. 3.7 with two major differences, the presence of an additional input to the controller and the absence of the delay block modeling the sample and hold process. The motivations for these differences will be given momentarily. In principle,

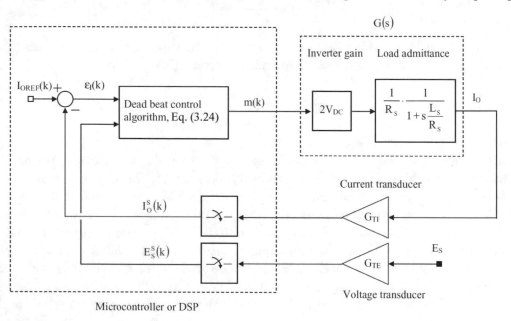

Figure 3.16: Block diagram of the dead-beat current control loop.

the dead-beat control strategy we are going to discuss is nothing but a particular application case of discrete time dynamic state feedback and direct pole allocation, and, as such, its formulation for our VSI model, as derived in Aside 1, can be obtained applying standard digital control theory. However, this *theoretical* approach is not what we are going to follow here. Instead, we will present a different derivation, completely equivalent to the theoretical one, but closer to the physical converter and modulator operation. We will discuss the equivalence of the two approaches in the Aside 5, but we feel like the physical one is somewhat easier to explain and better puts into evidence the merits and limitations of the predictive controller. For this reason, we chose to begin our discussion exactly from the *physical* approach.

Derivation of the Predictive Controller

The reasoning behind the physical approach to predictive current control is quite simple and can be explained referring to Fig. 3.17, representing an average model of the VSI and its load. At any given control iteration, we want to find the average inverter output voltage, $\overline{V}_{OC}$, that can make the average inductor current, $\overline{I}_O$, reach its reference by the end of the modulation period *following* the one when all the computations are performed. In other words, at instant $k \cdot T_S$

we perform the computation of the $\overline{V}_{OC}$ value that, once generated by the inverter, during the modulation period from $(k + 1) \cdot T_S$ to $(k + 2) \cdot T_S$, will make the average current equal to its reference at instant $(k + 2) \cdot T_S$. Please note that, doing so, the computation and modulation delays are inherently taken care of, and the controller dynamic response, as we are going to show, turns out to be equivalent to a pure two modulation period delay. Following this reasoning, the

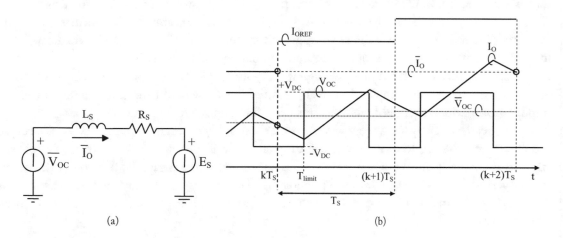

(a)   (b)

Figure 3.17: (a) Average equivalent circuit for the VSI of Fig. 3.1 and (b) average and instantaneous idealized waveforms for the inverter of Fig. 3.1.

control equation can be easily derived. Examining Fig. 3.17(a) the average inductor current at the instant $(k + 1)T_S$, is given by:

$$\bar{I}_O(k + 1) = \bar{I}_O(k) + \frac{T_S}{L_S} \cdot \left[\overline{V}_{OC}(k) - E_S(k) \right], \qquad (3.21)$$

where resistance R_S has been considered negligible, as it is often the case. Equation (3.21) simply expresses the physical fact that the current variation in an inductor is proportional to the integral of the applied inductor voltage. This voltage integral has been computed exploiting, once again, the Euler numerical integration rule. In other words, we are here considering again a zero order hold discrete time equivalent of the dynamic system represented in Fig. 3.17(a). Please note that, in (3.21) and according to Fig. 3.17(b), $\overline{V}_{OC}(k)$ indicates the average inverter voltage to be generated in the modulation period *following* the sampling instant $k \cdot T_S$, when all calculations are performed.

In principle, from (3.21) it would be possible to compute the $\overline{V}_{OC}(k)$ value required to make $\bar{I}_O(k + 1)$ equal to $I_{OREF}(k)$, thus achieving a one cycle delay dynamic response for the closed loop controlled system. In practice, since the computation of voltage $\overline{V}_{OC}(k)$ value occupies part of the modulation period, it is not possible to guarantee that, in all cases, the calculations will be over before the output voltage has to change its state from negative to positive, instant indicated by T_{limit} in Fig. 3.17(b). Please note that, in all cases when the average voltage to be applied is strongly positive, instant T_{limit} will be very close to instant $k \cdot T_S$, thus leaving very little time for computations.

The simplest way to solve these timing problems is to move the objective of the computation one step forward, i.e., instead of computing $\overline{V}_{OC}(k)$, we will now find the expression of $\overline{V}_{OC}(k+1)$. This will give us a whole modulation period to complete the calculations. Please note that we followed exactly the same approach when we modeled the calculation delay for the digital PI controller, considering it to be equal to one modulation period.

Also, similarly to the case of the digital PI, more sophisticated modeling approaches, taking into account the exact computation delay and duty-cycle update instant allocation within the modulation period are indeed possible, but we will not consider them here.

Instead, in Chapter 4, we will see how a thoroughly different organization of the control hardware and the implementation of the controller in a FPGA *circuit* rather than as a software algorithm, actually allows to reduce the computation delay to a negligible fraction of the modulation period. This way, the minimum possible small-signal response delay can be achieved, equal to a *half* of the modulation period. For now, all we can do is re-write (3.21) one step forward, thus getting

$$\bar{I}_O(k+2) = \bar{I}_O(k+1) + \frac{T_S}{L_S} \cdot \left[\overline{V}_{OC}(k+1) - E_S(k+1) \right]$$

$$= \bar{I}_O(k) + \frac{T_S}{L_S} \cdot \left[\overline{V}_{OC}(k+1) + \overline{V}_{OC}(k) - E_S(k+1) - E_S(k) \right], \qquad (3.22)$$

where $\bar{I}_O(k+1)$ has been replaced by its expression (3.21). Assuming now that voltage E_S is a slowly varying signal, as it is often the case, whose bandwidth is much lower than the modulation and sampling frequency, it is possible to consider $E_S(k+1) \cong E_S(k)$, thus obtaining the following dead-beat control equation

$$\overline{V}_{OC}(k+1) = -\overline{V}_{OC}(k) + \frac{L_S}{T_S} \cdot \left[\bar{I}_O(k+2) - \bar{I}_O(k) \right] + 2 \cdot E_S(k), \qquad (3.23)$$

where $\bar{I}_O(k+2)$ can be replaced by $I_{OREF}(k)$, the desired set-point. Equation (3.23) can be used to determine the duty-cycle, for the modulation period starting at instant $(k+1) \cdot T_S$, that will make the inductor current reach its reference at instant $(k+2) \cdot T_S$, with a two modulation period delay. If this holds, and indeed it does, application of (3.23) makes the closed-loop system dynamic response equivalent to a pure delay, i.e., guarantees a dead-beat control action.

As it is possible to observe, the application of (3.23) requires voltage E_S to be measured every sampling period, so that, differently from the PI current controller, the predictive controller, at least in this basic implementation, requires the sensing and analog to digital conversion not only of the regulated variable, i.e., the output current, but also of the phase output voltage.

Another, less evident, point regarding (3.23) is that, in general, the set-point for the average inverter output voltage it provides us with, will have to be correctly scaled down, so as to fit it to the digital pulse width modulator. The fitting is normally accomplished *normalizing* the output of the controller to the inverter voltage gain. In addition to this, the control equation has to be modified also to properly account for the transducer gains of both current and voltage sensors. It is easy to verify that an equivalent control equation, taking into account the transducer gains and voltage normalization, is the following:

$$m(k+1) = -m(k) + \frac{L_S}{T_S} \cdot \frac{1}{2 \cdot G_{TI} \cdot V_{DC}} \left[I_{OREF}^S(k) - I_O^S(k) \right]$$

$$+ 2 \cdot \frac{1}{2 \cdot G_{TE} \cdot V_{DC}} E_S^S(k), \tag{3.24}$$

where $m(k)$ is the modulating signal, input of the digital PWM and all variables are now *internal* variables, properly scaled down to fit to the microprocessor arithmetic unit. Please note that (3.24) also assumes that the modulating signal is bipolar, ranging between plus and minus one half of the modulator full scale input. As we did for the digital PI, this is assumed to be equal to unity, without any loss of generality. Under these assumptions, (3.24) can be turned into a control algorithm, to be programmed in the microcontroller or DSP unit.

An example of the predictive controller dynamic performance is shown in Fig. 3.18. In particular, Fig. 3.18(a) shows the reference tracking capability of the controller in the presence of a step change in the sinusoidal current reference. A more detailed view of the transient is shown in Fig. 3.18(b). Besides, in Fig. 3.18(c) the step response of the controller is simulated in the same operating conditions of the experimental set-up. The *experimental* verification of step response is shown in Fig. 3.18(d). As can be seen, despite the non ideal inverter behavior and the unavoidable parameter and model mismatches between the model and the experimental set-up, Fig. 3.18(d) is really very similar to Fig. 3.18(c).

It is interesting to compare these plots with those of Figs. 3.11 and, more fairly, 3.12. As can be seen, in spite of the calculation delay, the dynamic response of the dead-beat controller is faster than that obtained with the PI controller (even when no calculation delay is applied) and zero error reference tracking is resumed almost exactly after only two modulation periods from the first controller intervention.

A simple improvement of the presented control strategy makes it possible to derive an estimation equation that allows to save the measurement of the phase output voltage E_S. As in the control equation's case, the estimation equation can be derived by simple physical considerations, basically referring to (3.21). Indeed, re-writing (3.21) one step backward we get

$$\bar{I}_O(k) - \bar{I}_O(k-1) = \frac{T_S}{L_S} \cdot \left[V_{OC}(k-1) - E_S(k-1) \right], \tag{3.25}$$

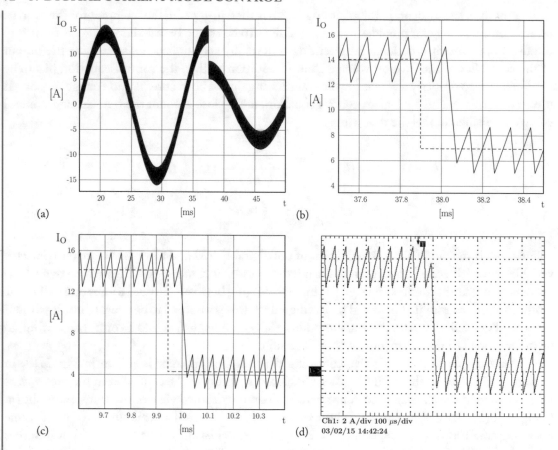

Figure 3.18: Simulation of the VSI with the predictive controller. The depicted variable is the VSI output current I_O. (a) Controller response to a step reference amplitude change. (b) Detail of previous figure. (c) Simulation of the controller's step response in the experimental set-up conditions. (d) Experimental verification of the controller's step response.

from which we can extract an estimation of $E_S(k-1)$. Simple manipulations of (3.25) yield

$$\hat{E}_S(k-1) = \overline{V}_{OC}(k-1) - \frac{L_S}{T_S} \cdot \left[\bar{I}_O(k) - \bar{I}_O(k-1)\right]. \tag{3.26}$$

Equation (3.26) represents the basic estimation equation for the predictive control of the VSI output current. It is typically possible to improve the quality of the estimation by using some form of interpolation or filtering, that can remove possible estimator instabilities.

Robustness of the Predictive Controller

The predictive controller derivation assumes that Equation (3.21) is a valid model of the VSI and its load. Although this is a generally solid assumption, in certain conditions the validity of (3.21) can be impaired. There can be at least two different reasons for this to happen, namely model mismatches and parameter uncertainties. An example of the former type is represented by not properly modeled circuit components (like R_S) that, assumed to be negligible, turn out to be comparable with the other circuit components.

The effect of model mismatches is normally very serious: since the control equation is based on a given system model, any deviation of the physical system from the model makes the controller interaction with the physical system unpredictable in its effects. Minor mismatches determine slight deviations from the expected dynamic performance, major ones can determine lightly damped closed loop responses or even make the system utterly unstable.

Aside 5. Derivation of the Predictive Controller Through Dynamic State Feedback

The VSI represented in Fig. 3.1 can be described in the state space by the following set of equations,

$$\begin{cases} \dot{x} = Ax + Bu \\ y = Cx + Du, \end{cases} \tag{A5.1}$$

which, as we recall from the discussion reported in Aside 1, can be used to relate average inverter electrical variables. In this case $x = [\bar{I}_O]$ is the state vector, $u = [\overline{V}_{OC}, \bar{E}_S]^T$ is the input vector, $y = [\bar{I}_O]$ is the output variable, and the state matrices are

$$A = [-R_S/L_S], \quad B = [1/L_S, -1/L_S], \quad C = [1], \quad D = [0, 0]. \tag{A5.2}$$

It is possible to derive a zero-order hold discrete time equivalent of (A5.1) considering the following system

$$\begin{cases} x(k+1) = \Phi x(k) + \Gamma u(k) \\ y(k) = Cx(k) + Du(k), \end{cases} \tag{A5.3}$$

where, by definition, $\Phi = e^{A \cdot T_S}$ and $\Gamma = (\Phi - I) \cdot A^{-1} \cdot B$. Computation of Φ and Γ yields

$$\Phi = e^{A \cdot T_S} = e^{-\frac{R_S}{L_S} \cdot T_S} \quad \overset{R_S \to 0}{\longrightarrow} \quad 1$$

$$\Gamma = \left[-\frac{e^{-\frac{R_S}{L_S} \cdot T_S} - 1}{R_S} \quad \frac{e^{-\frac{R_S}{L_S} \cdot T_S} - 1}{R_S} \right] \quad \overset{R_S \to 0}{\longrightarrow} \quad \left[\frac{T_S}{L_S} \quad -\frac{T_S}{L_S} \right], \tag{A5.4}$$

where both matrices have been calculated for the limit condition where the R_S value is negligible. Of course, this approximation is not strictly necessary to perform the following calculations and could be avoided. However, since we want to compare the results provided by the theoretical approach with those provided by the physical approach, we need to operate under the same conditions, which motivates the assumption of a negligible R_S value.

Given (A5.3), we can derive the predictive controller as a particular case of state feedback and pole placement. In order to show that, we may consider again (A5.3), rewriting the state equations explicitly. We get the following result:

$$\Sigma : \begin{cases} \bar{I}_O(k+1) = \bar{I}_O(k) + \dfrac{T_S}{L_S} \cdot \overline{V}_{OC}(k) - \dfrac{T_S}{L_S} \cdot \overline{E}_S(k) \\ y(k) = \bar{I}_O(k). \end{cases} \tag{A5.5}$$

Please note that the first equation in (A5.5) is exactly equal to (3.21). It is easy to verify that (A5.5) is exactly equivalent to the part of the following block diagram indicated by Σ. As can be seen, the block diagram includes the feedback controller as well. This schematic representation puts into evidence some interesting features of the considered discrete time system. In the first place, the diagram reveals how the $\overline{E}_S$ input can be considered an exogenous disturbance, whose compensation can be obtained by adding a suitable signal, ideally the $\overline{E}_S$ signal itself, to the control input $\overline{V}_{OC}$. As we will see in the following, the computation delay will make it impossible to get a perfect compensation of the disturbance, and, consequently, only partial compensation will be achieved. In addition, the block diagram shows how the feedback controller is itself a *dynamic system*. Differently from what is often done in state feedback applications, we are using here *dynamic state feedback* instead of a simple *static* feedback.

This is done to make the modeling of the computation delay more direct. In this case, referring to the diagram, we can represent the controller by means of the following state equation:

$$\overline{V}_{OC}(k+1) = K_2 \cdot \overline{V}_{OC}(k) + K_1 \cdot [I_{OREF}(k) - \bar{I}_O(k)], \tag{A5.6}$$

where, of course, the identity $\bar{I}_O = y$ has been used. The interconnection of Σ and the controller feedback generates a new, augmented, dynamic system, indicated by Σ_A. This is described by the following equations:

$$\Sigma_A : \begin{cases} \bar{I}_O(k+1) = \bar{I}_O(k) + \dfrac{T_S}{L_S} \cdot \overline{V}_{OC}(k) - \dfrac{T_S}{L_S} \cdot \overline{E}_S(k) \\ \overline{V}_{OC}(k+1) = K_2 \cdot \overline{V}_{OC}(k) + K_1 \cdot [I_{OREF}(k) - \bar{I}(k)] + K_3 \cdot \overline{E}_S(k) \\ y(k) = [1\ 0] \cdot \begin{bmatrix} \bar{I}_O(k) \\ \overline{V}_{OC}(k) \end{bmatrix}, \end{cases} \tag{A5.7}$$

which correspond to the state vector augmentation to $x_A = [\bar{I}_O \ \bar{V}_{OC}]^T$, to the new input vector $u_A = [\bar{E}_S \ I_{OREF}]^T$ and to the *approximated* compensation of the exogenous disturbance, governed by gain K_3. The corresponding state matrices are as follows:

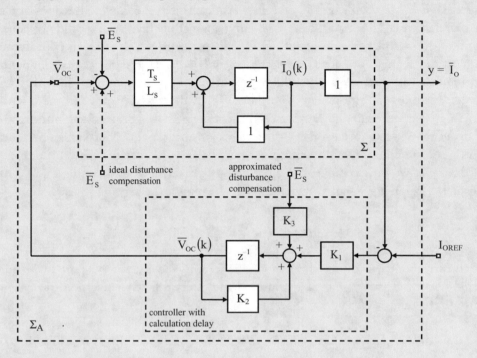

Figure A5.1: Block diagram of the discrete time models for the VSI and the dead-beat controller.

$$\Phi_A = \begin{bmatrix} 1 & \dfrac{T_S}{L_S} \\ -K_1 & K_2 \end{bmatrix}, \quad \Gamma_A = [\Gamma_{A1} \mid \Gamma_{A2}] = \begin{bmatrix} -\dfrac{T_S}{L_S} & \Big| & 0 \\ K_3 & \Big| & K_1 \end{bmatrix},$$

$$C_A = [1 \ 0], \quad D_A = [0 \ 0]. \tag{A5.8}$$

Based on (A5.8) it is possible to determine parameters K_1, K_2, and K_3 to get the desired pole allocation and disturbance compensation. It is easy to verify that choosing

$$K_1 = \frac{L_S}{T_S}, \quad K_2 = -1 \tag{A5.9}$$

the eigenvalues of Σ_A move to the origin of the complex plane. As is well known, this is a sufficient condition to achieve a dead-beat closed loop response from the controlled system.

Alternatively, the position of poles on the complex plane can be chosen to achieve a different closed loop behavior, for example one equivalent to that of a continuous time, first-order, stable system, characterized by any desired time constant. Indeed, with the direct discrete time design of the regulator, the designer has, in principle, complete freedom in choosing the preferred pole allocation. It is possible to demonstrate that the allocation of poles in the origin of the complex plane makes the closed loop system behavior very peculiar, bearing no similarity with any continuous time system's one. That is because the position of poles in the continuous time domain corresponding to the origin of the complex plane in the discrete time domain is minus infinity, which is not physically realizable, of course. In the discrete time domain instead, the allocation in the origin is perfectly realizable and determines the typical dead-beat closed loop behavior, i.e., the step response of the close loop system becomes equal to a linear combination of different order pure delays. In order to verify this property with our example, we now compute the closed loop transfer function between input I_{OREF} and output $\bar{I}_O$. Applying standard state feedback theorems and after simple calculations, we find

$$\frac{\bar{I}_O}{I_{OREF}}(z) = C_A \cdot (zI - \Phi_A)^{-1} \cdot \Gamma_{A2} = \frac{1}{z^2}, \tag{A5.10}$$

which corresponds, as expected, to a dynamic response equivalent to a pure two modulation period delay. Similarly, we can compute the closed loop transfer function from the disturbance to the output. We find

$$\frac{\bar{I}_O}{E_S}(z) = C_A \cdot (zI - \Phi_A)^{-1} \cdot \Gamma_{A1} = \frac{1}{z^2} \cdot \frac{T_S}{L_S} \cdot (-z - 1 + K_3). \tag{A5.11}$$

As can be seen, there is no value of K_3 that can guarantee a zero transfer function from disturbance to output. This is due to the fact that the compensation term of the controller equation (A5.7) is one step delayed with respect to the control output and, as such, is only approximated. Under these conditions, the best we can do is to minimize the transfer function (A5.11). It is easy to verify that the choice $K_3 = 2$ achieves this minimization. Rewriting (A5.11) in the time domain and imposing $K_3 = 2$ we find

$$\bar{I}_O(k) = \frac{T_S}{L_S} \cdot [-E_S(k - 1) + E_S(k - 2)], \tag{A5.12}$$

which, under the assumption of slowly varying E_S, guarantees the minimum disturbance effect of the output. Having determined the controller parameters K_1, K_2, and K_3, we are now ready to explicitly write the control equation, which turns out to be

$$\overline{V}_{OC}(k + 1) = -\overline{V}_{OC}(k) + \frac{L_S}{T_S} \cdot [I_{OREF}(k) - \bar{I}_O(k)] + 2 \cdot \overline{E}_S(k). \tag{A5.13}$$

As can be seen, (A5.13) is equal to (3.23).

To complete this theoretical discussion of the predictive controller, we need to add a final remark about the phase output voltage E_S estimation. The derivation presented here refers just to the basic predictive controller implementation, where E_S is assumed to be measured. If we want to consider the use of an output voltage estimator, additional care must be taken. The estimation equation can be directly obtained from the state variable $\bar{I}_O$ update equation (A5.5). However, using the estimated voltage $\hat{E}_S$ instead of the measured one in the control equation determines an increase of the order of the system, because $\hat{E}_S$ is a function of input and state variable values. As a general rule, this makes the dead-beat properties and stability of the controlled system more sensitive to model and parameter mismatches, reducing its robustness.

Parameter uncertainties, instead, are typically determined by construction tolerances or parameter value drifts, such as those due to varying physical or environmental operating conditions (current, temperature). Their effect on the dynamic performance of the closed loop system can be serious, ranging from different extents of performance degradation to system instability.

The formal analysis of model mismatches goes beyond the scope of our discussion, requiring a solid background in system identification theory. Instead, we can briefly discuss the effect of parameter uncertainty and provide an estimation of the predictive controller robustness to parameter variations. Considering (3.24) we see that several parameters contribute to the definition of the algorithm coefficients, each of them being a potential source of mismatch. To give an example of the analysis procedure we can apply to estimate the sensitivity of the controller to the mismatch we begin by referring, for simplicity, to (3.23), where the only parameter we need to take into account is inductor L_S. Of course, transducer gain or DC link voltage variations can be treated similarly. We can easily model errors or variations on parameter L_S distinguishing the value used in the VSI model from the one used in the predictive controller equation. In order to do that we can re-write (3.23) as follows:

$$\overline{V}_{OC}(k+1) = -\overline{V}_{OC}(k) + \frac{L_S \pm \Delta L_S}{T_S} \cdot \left[\bar{I}_O(k+2) - \bar{I}_O(k)\right] + 2 \cdot E_S(k), \qquad (3.27)$$

where parameter L_S has been replaced by $L_S \pm \Delta L_S$, thus putting into evidence the possible presence of an error, ΔL_S, implicitly defined as a positive quantity. The analysis of the impact of ΔL_S on the system's stability requires the computation of the system's eigenvalues. Referring to the procedure outlined in the Aside 5, we can immediately find the state matrix corresponding to (3.21) and (3.27). This turns out to be:

$$\Phi'_A = \begin{bmatrix} 1 & \frac{T_S}{L_S} \\ -\frac{L_S \pm \Delta L_S}{T_S} & -1 \end{bmatrix}. \qquad (3.28)$$

It is now immediate to find the eigenvalues of matrix Φ'_A. These are given by the following expression:

$$\lambda_{1,2} = \sqrt{\pm\frac{\Delta L_S}{L_S}} \qquad \Rightarrow \qquad |\lambda_{1,2}| = \sqrt{\frac{\Delta L_S}{L_S}}. \tag{3.29}$$

From (3.29) we see that the magnitude of the closed loop system's eigenvalues is limited to the square root of the relative error on L_S. This means that, unless a higher than 100% error is made on the estimation of L_S or, equivalently, unless a 100% variation of L_S takes place, due to changes in the operating conditions, the predictive controller will keep the system stable. Please note that, interestingly, this result is independent of the sampling frequency. Of course, even if instability requires bigger than unity eigenvalues, the good reference tracking properties of the predictive controller are likely to get lost, even for smaller than unity values of the relative error. We can visualize the effect of ΔL_S considering Fig. 3.19. Differently from previous figures, Fig. 3.19 shows the sampled current and its reference, to better put into evidence the effect of the parameter mismatch. We can immediately see how the presence of a mismatch determines an oscillatory response. We may also note that the undershoot and the damping factor are both affected by the amount of mismatch. With a 95% mismatch the response is lightly damped and barely acceptable for practical applications.

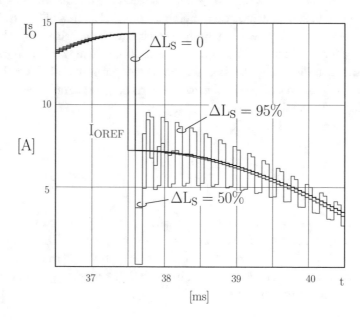

Figure 3.19: Simulation of the VSI with the predictive controller and different level of mismatch on parameter L_S. The figure shows the response to a step reference change of the VSI output current I_O.

As it might be expected, the robustness of the predictive controller to mismatches gets worse if the estimation of the phase voltage is used instead of its measurement. The analytical investigation of this case is a little more involved than the previous one, but still manageable with pencil and paper calculations. The procedure consists in writing the system (3.21), controller (3.23) and estimator (3.26) equations, either solving them using $\mathcal{Z}$-transform to find the reference to output transfer function, or, equivalently, arranging them to get the state matrix, and, finally, examining the characteristic polynomial of the system. Following the former procedure, we get:

$$\lambda\left(z\right)=z^{3}\pm3\cdot\frac{\Delta L_{S}}{L_{S}}\cdot z\pm2\cdot\frac{\Delta L_{S}}{L_{S}}.\tag{3.30}$$

It is now possible to plot the zeros of the characteristic polynomial, i.e., the closed loop system eigenvalues, as functions of ΔL_{S}. Considering only negative signs in (3.30), we find the results presented in Fig. 3.20. The figure shows that only a 20% error is allowed before system instability occurs. It is worth noting that this result is independent of the switching frequency since (3.30) is only a function of the mismatch error ΔL_{S}. Moreover, it is interesting to note that the unstable pole is at half of the sampling frequency, since it lies on the real axis. Instead, considering positive signs in (3.30) a similar situation can be found, where the minimum variation required to induce instability is somewhat higher than the previous one. Figure 3.20, therefore, represents the worst case condition.

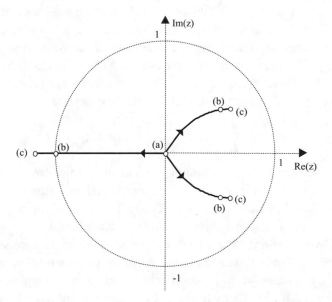

Figure 3.20: Plot of the closed loop system eigenvalues as functions of the parameter L_{S} mismatch. (a) $\Delta L_{S}=0$. (b) $\Delta L_{S}=0.2\cdot L_{S}$. (c) $\Delta L_{S}=0.3\cdot L_{S}$.

The results we have just found seem to completely undermine the practical applicability of the dead-beat current controller, especially when the output voltage estimation is considered. Luckily, this is hardly the case. The reason is that it is possible, with some modifications of the controller structure, as those suggested in [10] and [11], to strongly improve the controller robustness, making it perfectly apt to practical applications.

Effects of Converter Dead-Times

Converter dead-times are another non-ideal characteristic of the VSI that is not taken into account by the model the predictive controller is based on. In a certain sense, their presence can be considered a particular case of model mismatch. We know from Section 2.1.4 that the presence of dead-times implies a systematic error on the average voltage generated by the inverter. The error has an amplitude that depends directly on the ratio between dead-time duration and modulation period and a sign that depends on the load current sign. As we did in Section 2.3.1, we can model the dead-times' effect as a square wave disturbance having a relatively small amplitude (roughly a few percent of the DC link voltage) and opposite phase with respect to the load current. We can consider this disturbance as an undesired component that is summed, at the system input, to the average voltage requested by the current control algorithm.

As such, the disturbance should be, at least partially, rejected by the current controller. The effectiveness of the input disturbance rejection capability depends on the low frequency gain the controller is able to determine for the closed loop system. And here is where the dead-beat controller shows another weak point. We have seen how the dead-beat action tends to get from the closed loop plant a dynamic response that is close to a pure delay. Unfortunately, this implies a very poor rejection capability for any input disturbance. To clarify this point we can again refer to the Aside 5, Fig. A5.1, and compute the closed loop transfer function from the exogenous disturbance $\overline{E}_S$ to the output $\overline{I}_O$. Indeed, this is the transfer function experienced by the dead-time induced voltage disturbance as well. Simple calculations yield:

$$\frac{\overline{I}_O}{\overline{E}_S}(z) = \frac{1}{z^2} \cdot \frac{T_S}{L_S} \cdot (z+1), \tag{3.31}$$

which means that the output signal is proportional to the sum of two differently delayed input samples. In terms of disturbance rejection the result is rather disappointing. Plotting the frequency response of (3.31) we find that it is practically flat from zero up to the Nyquist frequency, i.e., there is no rejection of the average inverter voltage disturbance.

Consequently, we cannot expect the dead-beat controller to compensate the dead-time effects. This means that, unless some external, additional compensation strategy is adopted, a certain amount of current distortion is likely to be encountered. A typical example is shown in Fig. 3.21, where the sampled output current and its reference are shown. We can clearly see the double effect of uncompensated dead-times: (i) a systematic amplitude error and (ii) a crossover distortion phenomenon. The reason for the former is an obvious consequence of the negligible rejection capability of the dead-beat controller. The latter instead is due to the sign inversion of the voltage disturbance, taking place at the moment of current zero crossing, that the controller tries to compensate. We can also note how, because of the current ripple amplitude, the sign of the voltage disturbance is not stable around the zero crossing instant, which makes the transient duration longer. It is worth noting that Fig. 3.21 is obtained assuming that the duration of the dead-time is about 5% of the modulation period. This is an exaggerated value, that is used in the simulation on purpose, just to magnify the current distortion. In practical situations, dead-times are in the range between 1% and 2% of the modulation period and the overall effect on the converter output current is accordingly smaller.

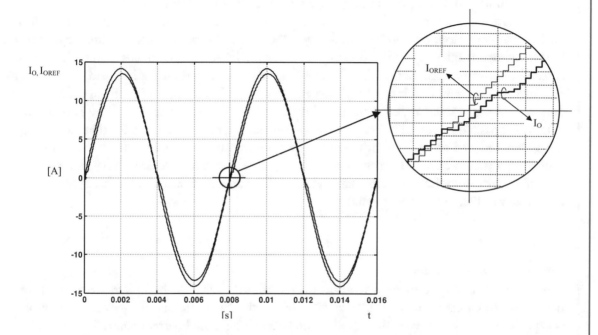

Figure 3.21: Simulation of the VSI with the predictive controller and dead-times. The dead-times' effects are: (i) crossover distortion, visible in the inset and (ii) steady-state error between the amplitude of the converter current and its reference. The sampled output current is shown instead of the instantaneous one to eliminate the ripple, that can mask the error. Dead-times are set to 5% of the modulation period, just to magnify the effect.

There are different possible methods to compensate the dead-time induced output current distortion, that are sometimes used also in conjunction with regulators, like the PI controller, that present a significant low frequency rejection capability. That is because such regulators are nonetheless exposed to crossover distortion which can be unacceptable for some applications, like for example high quality electrical drives. In the case of the dead-beat controller some form of compensation is instead mandatory. Compensation methods can be divided into: (i) closed loop or on-line and (ii) open loop or off-line. The best performance is offered by closed loop dead-time compensation, that requires, however, the measurement of the actual inverter average output voltage. Its comparison with the voltage set-point provided to the modulator gives sign and amplitude of the dead-time induced average voltage error, that can therefore be compensated with minimum delay, simply by summing to the set-point for the following modulation period the opposite of the measured error. The need for measuring the typically high output inverter voltage requires the use of particular care. The estimation of the inverter average output voltage is normally done by measuring the duration of the voltage-high and voltage-low parts of the modulation period, i.e., by computing the actual, effective output duty-cycle.

However, much more often, off-line compensation strategies are used. These offer a lower quality compensation, but can be completely embedded in the modulation routine programmed in the microcontroller (or DSP), requiring no measure. The-off line compensation of dead-times is based on a worst case estimation of the dead-time duration and on the knowledge of the sign of the output current, that is normally inferred from the reference signal (not from the measured output current, to avoid any complication due to the high frequency ripple). Given both of these data, it is possible to add to the output voltage set-point a compensation term that balances the dead-time induced error. The method normally requires some tuning, in order to avoid under or over compensation effects. The results are normally quite satisfactory, unless a very high precision is required by the application, allowing to eliminate the amplitude error and to strongly attenuate the crossover distortion phenomenon.

Comparison with PI Controller

A final remark is needed to summarize the main features of the dead-beat predictive current controller and to compare its performance with that of the PI controller. The predictive controller is capable of a very fast dynamic response, the best among digital current controllers and clearly superior to that achievable by any digital PI controller. Therefore, it is very suited to those applications of VSIs where the capability of tracking rapidly variable current reference signals is required. Examples of these applications can be considered the active power filters and the high performance adjustable speed drives. On the other hand, the predictive controller, at least in its basic implementation, requires the measurement of the load voltage, which generally complicates the hardware needed for its implementation well beyond what is required by a PI controller. We have also seen how estimation techniques can be employed to avoid the voltage measurement, but we need to point out that: (i) the estimation makes the controller more sensitive to model and parameter mismatches and (ii) the dynamic performance is worsened, although it usually remains better than that of a conventional PI regulator. Finally, we have seen how the compensation of dead-times is practically mandatory for the dead-beat controller, that has no inherent integral action, while it may not be required by the PI, unless very low distortion current waveforms are required by the application. Moreover, the sensitivity of the predictive controller to measurement noise is surely higher than that of the PI controller, which calls for particular care in the design of the signal conditioning circuitry.

REFERENCES

[1] A.V. Oppenheim, R.W. Schafer, J.R. Buck, "Discrete Time Signal Processing," Second Edition, Prentice Hall, 1999, ISBN 0–13-083443-2. 36, 41, 63

[2] K. Ogata, "Discrete time control systems," Prentice Hall, 1987, ISBN 0–13-216102-8. 39, 41, 50

[3] K.J. Astrom, B. Wittenmark, "Computer-Controlled Systems: Theory and Design," Third Edition, Prentice Hall, 1997, ISBN 0–13-314899-8. 41, 50

[4] A.V. Peterchev, S.R. Sanders, "Quantization resolution and limit cycling in digitally controlled PWM converters," *IEEE Transactions on Power Electronics*, Vol. 18, No. 1, January 2003, pp. 301–308. DOI: 10.1109/TPEL.2002.807092. 45

[5] H. Peng, D. Maksimovic, A. Prodic, E. Alarcon, "Modeling of Quantization Effects in Digitally Controlled DC-DC Converters," *IEEE Transactions on Power Electronics*, Vol. 22, No. 1, January 2007, pp. 208-215. DOI: 10.1109/TPEL.2006.886602. 45, 46

[6] D.M. Van de Sype, K. De Gusseme, A.R. Van den Bossche, J.A. Melkebeek, "Small-signal z-domain analysis of digitally controlled converters," Power Electronics Specialists Conference (PESC), 20–25 June, 2004, pp. 4299–4305. DOI: 10.1109/PESC.2004.1354761. 60

[7] G.C. Verghese, M.E. Elbuluk, and J.G. Kassakian, "A general approach to sampled-data modeling for power electronic circuits," *IEEE Transactions on Power Electronics*, Vol. 1, No. 2, April 1986, pp. 76–89. DOI: 10.1109/TPEL.1986.4766286. 60

[8] J. Kassakian, G. Verghese, M. Schlecht, "Principles of Power Electronics," 1991, Addison Wesley, ISBN 02010–9689-7. 60

[9] B.K. Kuo, "Digital Control Systems," SRL Publishing Company, Champaign, Illinois, 1977. 66

[10] L. Malesani, P. Mattavelli, S. Buso, "Robust Dead-Beat Current Control for PWM Rectifiers and Active Filters," *IEEE Transactions on Industry Applications*, Vol. 35, No. 3, May/June 1999, pp. 613–620. DOI: 10.1109/28.767012. 67, 80

[11] G.H. Bode, P.C. Loh, M.J. Newman, D.G. Holmes "An improved robust predictive current regulation algorithm," *IEEE Transactions on Industry Applications*, Vol. 41, No. 6, November/December 2005, pp. 1720–1733. DOI: 10.1109/TIA.2005.858324. 67, 80

CHAPTER 4

Multi-Sampled Current Controllers

As mentioned in Section 2.2.4, oversampling may represent an effective provision to mitigate the digital PWM intrinsic delay and widen the current control-loop bandwidth.

The implementation of multi-sampled current controllers, where the current error is acquired multiple times in a modulation period, is not exactly straightforward. Indeed, in order to get relatively high oversampling factors, e.g., higher than ten, no conventional control hardware, like a microcontroller or a DSP, can be used. Built in AD converters are, indeed, too slow, but, most of all, it is the computation time of the control algorithm that limits the sampling rate to relatively low frequencies.

Instead, adopting a different hardware organization, it is actually possible to develop effective, high performance multi-sampled controllers. The enabling technologies, from this standpoint, are represented by:

- large bandwidth AD converters, capable of operating at several mega sample per second (M*Sample/s*) speeds;

- programmable digital hardware, i.e., FPGA chips, that allow to minimize the computation time of the control algorithm, thus keeping pace with the ADC.

Both are currently available at relatively low cost, so that the realization of multi-sampled controllers is nowadays as affordable as it has never been before. On the other hand, since the diffusion of ready-to-use control platforms is still quite limited, custom-designed control hardware needs to be assembled and validated *before* the current controller development can even start, which may imply somewhat longer development times, with respect to conventional solutions.

Nevertheless, it is obviously very interesting to investigate how to fully exploit oversampling in the implementation of current controllers, because they represent, it is worth remarking, the backbone of a large variety of more complex applications of VSIs. As will be clarified in Chapter 6, maximizing the current loop performance can positively impact on other external control loops, increasing their stability margins and robustness. For this reason, in this chapter, we would like to discuss the multi-sampled implementation of the basic current controllers for voltage source inverters that have been previously presented.

The objective of the discussion is twofold: (i) to identify the achievable performance limits for each of the presented techniques and (ii) to highlight the advantages of the multi-sampled implementations with respect to their conventional counterparts.

In order to do so, we will have, first of all, to exactly define the control hardware characteristics, because these are no longer irrelevant in determining the controller performance. This makes a big difference with respect to software defined controllers, that are, by definition, portable, i.e., compatible with *any* programmable device offering the required computational capacity.

Our discussion will take three different multi-sampled, high-performance current controller implementations into consideration, namely:

1. the oversampled proportional integral (PI) controller;

2. the oversampled implementation of the predictive current controller, [1];

3. the fully digital, fixed frequency implementation of the hysteresis current controller [2].

To simplify the explanation and make the three solutions fairly comparable from different standpoints, they are supposed to be designed referring to the same sensing and signal conditioning circuitry, the same high performance, 12 bit analog to digital converter (ADC) and to be synthesized on the same FPGA chip. The basic features of the considered FPGA and ADC hardware are summarized in Table 4.1.

Table 4.1: FPGA and ADC chip characteristics

Component	Model	Parameter	Value
FPGA	Spartan-6 LX45	Slices DSP48s Max. clock frequency Arithmetic resolution	6822 58 200 MHz 16 bit
ADC	AD9226	Max. sampling rate Full scale range Resolution Latency @ 30MHz	65 M*Sample / s* 3V 12 bit 0.286 µs

In all cases, signal processing is performed, by suitable FPGA synthesized algorithm computation circuits, on 16 bits, with fixed point arithmetic. As we will see, the 16 bit resolution is sufficient to prevent arithmetic quantization issues.

As far as the test bench converter topology is concerned, we keep referring to the simple VSI of Fig. 2.1, with the parameters reported in Table 2.1.

4.1   OVERSAMPLED PI CURRENT CONTROLLER

The hardware configuration of the oversampled PI controller is schematically shown in Fig. 4.1. As can be seen, the current signal is sampled, and subsequently processed, at the occurrence of

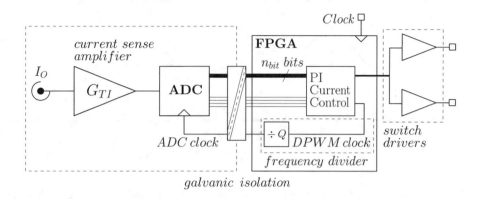

Figure 4.1: Hardware organization of the oversampled PI current controller.

an ADC clock pulse, that is derived from the DPWM clock by frequency division. The current reference is internally generated; its waveform can be selected through a dedicated input port, not shown in the figure. The current error processing results in the typical steady-state signals shown in Fig. 4.2, where the considered triangular DPWM implementation is also visible. The DPWM clock frequency is an *even* integer multiple of the switching frequency, so that each modulation period is divided into $2M$ DPWM clock periods, with

$$M = \frac{f_{DPWM}}{2\,f_S}. \tag{4.1}$$

The current error is calculated every Q DPWM clock cycles, with Q chosen among the integer sub-multiples of M, i.e., such that $M/Q = P$, $P \in \mathbb{N}$. As a result, the oversampling factor of the controller, N, can be defined as

$$N = 2\,\frac{M}{Q}. \tag{4.2}$$

It is typically lower than M, as much as required by the ADC conversion speed.

The PI controller gains can be easily selected, just as we did for the conventional implementation in Section 3.2, imposing the desired crossover frequency and phase margin, essentially using the same procedure outlined in Aside 2. Minor adjustments are required to take into account the different modulator gain, now equal to M and its lower delay.

As we have seen in Section 2.2.4, the small-signal delay of the DPWM with symmetrical, triangular carrier can be pretty well approximated by $T_S/2$, which exactly corresponds to a ZOH

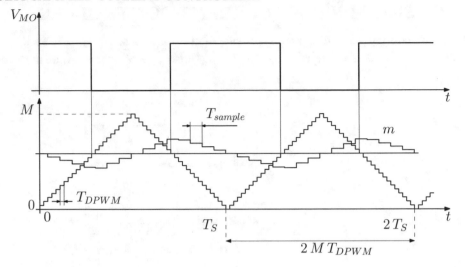

Figure 4.2: Triangular digital pulse width modulator operation with a multi-sampled proportional integral current controller. Note that the c_{pk} amplitude has been set equal to M.

delay for the applied switching frequency. In our case, however, thanks to oversampling, the delay is reduced by a factor N with respect to its conventional, single-sample implementation [5].

Besides, the algorithm calculation time is practically negligible. At the considered FPGA clock frequency (see Table 4.2) it roughly amounts to 33 ns, representing just the 0.067% of the modulation period, i.e., 3.3% of the adopted sampling period. Therefore, the design procedure adopted for the analog PI regulator, where no calculation delay was taken into account is still applicable.

As a result, the current loop bandwidth could be pushed closer to its theoretical limit, determined by the converter output impedance, exactly as if we were dealing with an analog PI controller, with negligible PWM or calculation delay. However, in order to facilitate the comparison with the results illustrated in previous chapters, in our design of the oversampled current controller, the crossover frequency f_{CL} was set to $f_S/6$, with a 60° phase margin, exactly as we did in Aside 2 designing the analog PI controller. The parameters considered for controller implementation are listed in Table 4.2. It is interesting to compare the multi-sampled PI specifications with the results reported in Table 3.2. We can recognize that the oversampled controller outperforms even the best results achieved by an optimized conventional PI with minimized computational delay.

Referring to the block diagram of Fig. 3.9, it is easy to verify that W_{IPI}, the closed loop transfer function between the current reference and the inverter output current in the continuous

Table 4.2: PI controller parameters

Parameter	Symbol	Value	
Clock frequency	f_{clock}	30	MHz
DPWM resolution	M	750	
Oversampling rate	N	50	
ADC resolution	n_{bit}	12	*bit*
Current sensor gain	G_{TI}	100	mV/A
Current sensor FSR	I_{MAX}	± 30	A

time domain, is given, under the same assumptions of Section 3.2.1, by the following expression:

$$W_{I_{PI}}(s) = \frac{I_O(s)}{I_{OREF}(s)} = \frac{PWM(s)\,G(s)\,PI(s)}{1 + PWM(s)\,G(s)\,PI(s)}, \tag{4.3}$$

where

$$PWM(s) = \frac{1}{M} \cdot e^{-\frac{T_S}{2N}};$$

$$G(s) = \frac{G_{TI}}{FSR} \frac{1}{R_S} \frac{2\,V_{DC}}{1 + s\frac{L_S}{R_S}}; \tag{4.4}$$

$$PI(s) = K_P + \frac{K_I}{s}.$$

Equation (4.3) is important to estimate the equivalent small-signal delay of the closed-loop current controller.

Aside 6. LCO Containment in the PI Controller

As was explained in Section 3.1.3, closing a control loop around a DPWM may cause the occurrence of limit cycle oscillations (LCOs), even when oversampling is not considered [3]. The analytical derivation of existence conditions for LCOs and of mathematically solid design criteria for LCO free, multi-sampled controllers is relatively involved; as an example, the reader can find a detailed analysis of this phenomenon in [4], where the case of oversampled voltage regulation loops in buck converters is taken into account.

Without introducing excessive mathematical complications, we found that a simplified reasoning can effectively prevent the occurrence of *disturbing* LCOs, although there is no proof it will be *always* as effective as in the case here presented. It is important to underline that LCOs whose amplitude is such that no significant perturbation of the regulated current can be observed, have a negligible impact on the controller performance. This is what typically happens when a relatively large resolution is available for both the DPWM and the ADC. The following approximate reasoning allows us to better define the DPWM resolution required to make the LCO practically undetectable.

Let's suppose a one LSB wide perturbation at half the switching frequency is superimposed to the converter duty-cycle, i.e., summed at the input of the DPWM, while the converter is operating in the steady-state. The perturbation can be considered to have *negligible consequence* on the current controller operation if it generates an *average* current variation, in a switching period, that is *lower than half the LSB* of the current error sample numerical representation, which we denote with q_{ADC}. Indeed, such a current variation will not trigger any short term correction of the *average* modulating signal level, because the PI controller will not modify its integral output in the immediately following switching period. If we consider that the modulating signal update instants are synchronized with the DPWM counter by the hardware, we have to conclude that the duty-cycle perturbation will not propagate through the control loop.

Basically, this nice outcome can be achieved if we design the DPWM so that:

$$f_{DPWM} \geq \frac{2 V_{DC}}{L_S q_{ADC}} = \frac{2^{n_{bit}} V_{DC}}{L_S |I_{MAX}|}. \tag{A6.1}$$

where I_{MAX} represents the full-scale range in [A] of the current sensor and n_{bit} is the ADC resolution. Clearly, (A6.1) does not represent a necessary nor a sufficient condition to prevent LCOs, because larger in amplitude, and/or longer in period, duty-cycle oscillations can still be generated, depending on several additional parameters, like the controller gains and the load applied to the converter.

Adopting (A6.1) as a design criterion, however, guarantees that enough resolution is available in the DPWM to prevent the most common LCO condition, determined by the non existence of a DC operating point for the modulator, from perturbing the current controller, generating a cycle-by-cycle instability. In general, this makes the impact of all the possible LCOs on the regulated current practically negligible. It is worth noting, finally, that this result is only made possible by the adoption of a FPGA synthesized control circuit. Indeed, a conventional DSP or microcontroller can hardly offer the required DPWM resolution.

4.1.1   SMALL-SIGNAL FREQUENCY RESPONSE

Measuring the small-signal response of a controller is a very simple and effective way to get a reliable estimation of its small-signal delay characteristics, in particular of the phase lag that it will introduce while tracking a sinusoidal signal of a given frequency.

As we will see in Chapter 6, outer control loops can be closed around the current control one. Their design is directly and heavily affected by the small-signal characteristics of the inner loop. Indeed, the phase lag introduced by the current controller determines the phase margin of the outer loop at its desired cross-over frequency and, consequently, its achievable bandwidth and overall stability margin. That is why we need to be particularly careful in determining the phase lag introduced by a current controller and why it is highly desirable for this figure to be as small as possible.

For linear controllers, like the multi-sampled PI and the predictive current controller, an estimation of the small-signal frequency response can be found analytically, although under some simplifying assumptions, considering the closed-loop continuous time transfer functions of the current control loop, like (4.3). For nonlinear controllers, like the hysteresis current controller, it is not possible to define any transfer function. Therefore, only numerical simulation or experimental measurements can give an estimation of the small-signal frequency response.

In all cases, experimental measurements must be performed and compared to analytical (or simulation) results, so as to provide the required verification. To this purpose, small (i.e., lower than 5% in relative terms) sinusoidal current reference perturbations can be injected into the control loop, simply summing the perturbing signal to a DC reference signal. The resulting converter output current can be easily measured and analyzed with the Discrete Fourier Transform (DFT). Considering a sufficiently large time span, it is possible to determine, with adequate resolution and precision, the phase and magnitude of the injected perturbation effect. Repeating the procedure on a predefined set of frequencies allows to derive a graph like the one of Fig. 4.3, where the measurement results are compared to the analytically calculated ones, resulting from (4.3).

The considered frequency span extends from 10 Hz to 3.5 kHz, because this is the range where the crossover frequency of the outer control loop is expected to lie. As can be seen, the experiments confirm the good quality of the closed loop model (4.3), especially considering that the magnitude scale is very fine. Indeed, the maximum magnitude deviation from the analytical model is lower than 1 dB. As far as the phase lag is concerned, we can notice how this is consistent with the chosen cross-over frequency for the current controller.

4.2   OVERSAMPLED PREDICTIVE CURRENT CONTROLLER

The hardware arrangement of the predictive current controller is shown in Fig. 4.4. As can be seen, the circuit is very similar to the PI controller's one, because the current signal is, once again, sampled and subsequently processed at the occurrence of an ADC clock pulse, that is derived by frequency division from the DPWM clock. A second ADC channel is used to acquire, at the same clock pulse, a sample of the inverter phase output voltage, E_S. Differently from the PI controller, in this case the current error is actually *down sampled*, so that the control circuit only uses the *average* current error samples, that are available only *twice* per modulation period.

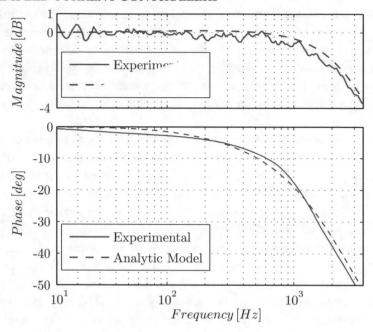

Figure 4.3: Multi-sampled PI controller small-signal frequency response in magnitude (upper plot) and phase (lower plot) according to analysis (dashed), experiments (solid).

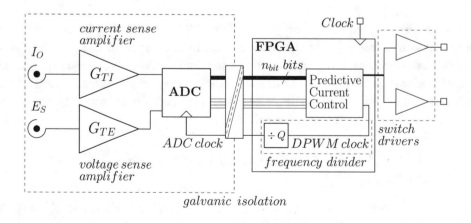

Figure 4.4: Hardware organization required by the oversampled predictive current controller.

The other samples are not simply discarded. Instead, they are passed to a trigger sub-circuit, whose purpose is to detect large-signal variations of the current reference, so as to allow the dead-beat controller to react to these perturbations with minimum delay, i.e., in just *one sampling period*.

Independently from that, the controller generates a new duty-cycle every *half* a switching period, implementing the following algorithm:

$$m(k) = \underbrace{\frac{L_S \, f_S}{V_{DC}}}_{k_\varepsilon} \cdot \overbrace{\frac{I_{OREF}^S(k) - I_O^S(k)}{G_{TI}}}^{\varepsilon_I} + \frac{E_S^S(k)}{2 \, V_{DC} \, G_{TE}} + \frac{1}{2}. \qquad (4.5)$$

Equation (4.5) is obtained considering the same first order inverter model we took into account in Chapter 3 and expressed by (3.21). It is therefore worth comparing it to (3.24) to see how the conventional controller is modified into the higher performance version discussed here.

The fundamental difference between the two implementations is given by the algorithm calculation time, which is thoroughly neglected in (4.5), while it determines a full modulation period delay in the original formulation. Indeed, as can be seen, (4.5) determines the duty-cycle for the kth control period based on the most recently acquired average current and output voltage samples, i.e., belonging to the same k^{th} time interval. This is possible thanks to the FPGA implementation of the controller. Because the algorithm is very simple, the FPGA circuit is able to calculate it in a single clock cycle, which is, in any case, a negligible amount of time when compared to the switching, or even to the sampling period (see Section 4.1). As a result, the modulating signal can be updated almost instantaneously, just as soon as new average current and output voltage samples are available.

But there is another important difference between the two implementations, that is the number of modulating signal updates per modulation period. In the case of (4.5), this is 2, i.e., the controller operates in double update mode, while, in the case of the original implementation presented in Chapter 3, just one update per modulation period has been obtained.

These differences guarantee a much higher performance to the multi-sampled dead-beat controller, as we will see in the following. A schematic view of the controller operation in the steady state is given by Fig. 4.5.

As usual, there are some drawbacks in this solution as well, the major one being the need for LCO limiting provisions. The algorithm is indeed exposed to limit cycle oscillation phenomena, as any controller driving a digital pulse width modulator. These can be caused, primarily, by:

1. insufficient resolution in the DPWM;

2. incorrect ratio between DPWM clock and sampling frequencies.

Each of these causes are discussed, and the appropriate countermeasures presented, in the Aside 7.

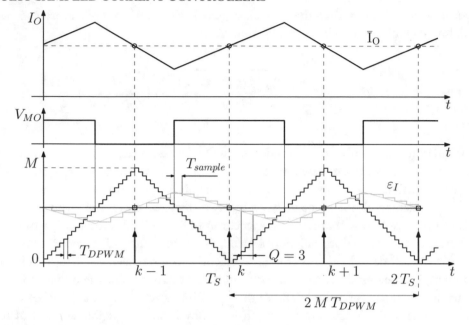

Figure 4.5: Basic multi-sampled predictive controller operation: the controller acquires $N \geq 2$ samples per switching period, but only those taken in the clock periods marked by the arrows are considered for duty-cycle updating. The quantity ε_I represents the current error.

Aside 7. LCO Containment in the Predictive Controller

As mentioned in Section 4.2, the oversampled predictive controller is exposed to LCO phenomena because of either an insufficient resolution in the DPWM or an incorrect ratio between DPWM clock and sampling frequencies. Let's analyze both problems in more detail.

DPWM Resolution

As mentioned earlier, the DPWM generates limit cycle oscillations any time the target duty-cycle does not match one of the physically realizable ones, corresponding to one of the possible counter values. Like in the PI controller's case, a simple condition to prevent the occurrence of persistent high frequency LCOs can be derived for the predictive controller as well. Considering Fig. 4.5 once again, it is possible to see that the minimum applicable duty-cycle correction is equal to $1/M$. If we perturb the steady state by applying such a correction, after half a modulation period, an average current variation will develop. In order not to trigger a persistent oscillation, such variation has to be undetectable by the controller, i.e., lower or equal than the current error least significant bit value, q_{ADC}. Assuming *uniform* quantization and *unsigned* representation for the current samples, this reasoning yields the following inequalities:

$$\frac{1}{M} \cdot \frac{2}{k_\varepsilon} \leq q_{ADC} \quad \Leftrightarrow \quad f_{DPWM} \geq \frac{V_{DC}}{L_S} \frac{2^{n_{bit}}}{|I_{MAX}|}, \tag{A7.1}$$

where I_{MAX} represents the full scale range of the current sensor. With the controller parameter values listed in Table 4.3, condition (A7.1) requires a 28.44 MHz minimum clock frequency. For this reason, $f_{clock} = 30$ MHz has been considered in the controller implementation discussed hereafter (as well as in the PI controller's case).

Table 4.3: Dead-beat controller parameters

Parameter	Symbol	Value	
Clock frequency	f_{clock}	30	MHz
DPWM resolution	M	750	
Oversampling rate	N	50	
ADC resolution	n_{bit}	12	*bit*
Current sensor gain	G_{TI}	100	mV/A
Voltage sensor gain	G_{TE}	15	mV/V
Current sensor FSR	I_{MAX}	±30	A

DPWM Clock and Sampling Frequency

Even if the controller hardware is designed to comply with (A7.1), a LCO free operation of the algorithm can be obtained only if two additional *necessary* conditions are satisfied: (i) the sampling process is synchronized with the modulator and (ii) an even number of current error samples is taken in each modulation period. Indeed, only if these conditions are met, in the steady state, both the current error zero crossings of each modulation period will take place at the correct sampling instants and, consequently, undesired transients will be avoided determining a smooth operation as the one shown in Fig. 4.5.

Synchronization is inherently ensured by the hardware organization of the controller of Fig. 4.4, because the ADC sampling clock is derived from the DPWM clock. To meet the second condition, instead, the following design constraint must be satisfied:

$$2M = N \cdot Q \quad N, M, Q \in \mathbb{N}, \ M \geq N \tag{A7.2}$$

which, as can be seen, is just a re-formulation of (4.2). This equation means that not *any* oversampling factor is compatible with a given DPWM resolution: in order to prevent LCO occurrence, the two must be chosen coordinately.

Once conditions (A7.1) and (A7.2) are satisfied, the controller design can be considered complete, although there are a few other minor details to take care of,[a] about which the interested reader can find more information in [1].

[a]We refer here to the modulator logic, the off-line compensation of dead-times, and to above-mentioned transient detection sub-circuit.

4.2.1 CLOSED-LOOP TRANSFER FUNCTION DERIVATION

Provided there are *no* parameter mismatches and *unmodeled* dynamics, like, e.g., the inductor equivalent series resistance, ESR, in the internal model the controller uses to regulate the current, the closed-loop small-signal response of (4.5) is exactly equivalent to an ideal *half a switching period* delay (i.e., to a dead-beat response). Indeed, the closed-loop transfer function between the current reference and the converter current sample sequences is given by:

$$W_{I_{DB}}(z) = \frac{I_O(z)}{I_{OREF}(z)} = \frac{G(z)H(z)}{1 + G(z)H(z)} = z^{-1}, \tag{4.6}$$

where

$$G(z) = \frac{V_{DC}}{L_S\,f_S}\frac{z^{-1}}{1-z^{-1}}; \tag{4.7}$$

$$H(z) = \frac{L_S\,f_S}{V_{DC}}.$$

Please note that, in deriving (4.7), an ideal converter impedance has been considered, with no ESR, i.e., with $R_S = 0$. We can now appreciate, comparing (4.6) to (A5.10) and considering that the sampling period is now $T_S/2$, how the oversampled implementation allows to reduce the small-signal response delay to *one fourth* with respect to the original solution. From (4.6) we can also immediately derive the equivalent, continuous time transfer function of the closed loop, oversampled predictive current controller. This is given by:

$$W_{I_{DB}}(s) = e^{-\frac{sT_S}{2}}, \tag{4.8}$$

that can be directly compared to (4.3). However, all the nice properties of the dead-beat controller need to be tested against model mismatches and some, unavoidable, parametric uncertainty, just like we did for the conventional implementation in Section 3.2.7. Because there is no intrinsic difference in the controller structure with respect to the conventional implementation, the same good robustness to parameter mismatches is obtained for the multi-sampled implementation as well.

4.2.2 SMALL-SIGNAL FREQUENCY RESPONSE

As we have done for the PI controller in Section 4.1.1, it is possible to measure the small-signal frequency response of the oversampled predictive current controller as well. The result is shown in Fig. 4.6, where measured data are compared to the analytical model (4.8). It is possible to

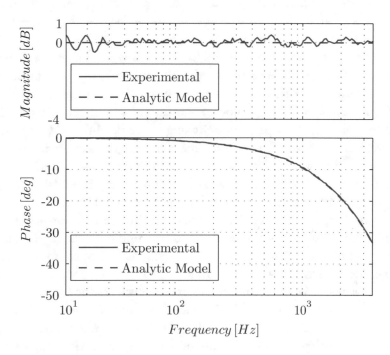

Figure 4.6: Multi-sampled predictive controller small-signal frequency response in magnitude (upper plot) and phase (lower plot) according to: analysis (dashed) and experiments (solid).

see how the experiments match the analytical results pretty well. It is also possible to compare these with Fig. 4.3 and to verify that both the magnitude and the phase response of the predictive controller is significantly better than the PI's. Indeed, the magnitude response is practically flat in the considered frequency span, while the phase lag is much lower. For example, at 3 kHz, the phase lag reduces from 49° to 28°.

It is important to underline that both the multi-sampled PI and the predictive controller really outperform their conventional counterparts, as can be easily verified using the models of Chapter 3, i.e., the one represented in Fig. 3.9, for the PI controller, and that corresponding to (A5.10), for the predictive one.

4.3   DIGITAL, FIXED FREQUENCY HYSTERESIS CURRENT CONTROLLER

As we have briefly discussed in Section 2.3.2, the hysteresis current control is, in principle, capable of excellent steady-state and dynamic performance. However, it normally implies variable switching frequency operation, because, in several applications, neither the converter's output voltage nor the current reference signal are perfectly constant.

Frequency stabilization is possible and has been proven to be quite effective in making the controller, from the spectral performance standpoint, practically equivalent to a PWM-based one, especially when digital hardware is used to the purpose. The resulting controller implementation is, accordingly, mixed signal. In addition to the current sensing circuitry, analog comparators are employed to determine the optimum switching instants, while DACs are needed to provide adjustable hysteresis thresholds. This exposes the controller to the uncertainties typical of analog circuits and related to offsets, drifts, tolerances, and aging effects. As discussed in [2], however, once a controller hardware of the type we are here considering is available, the implementation of a *fully digital* hysteresis current controller becomes perfectly feasible.

The hardware organization of the controller, in this case, is shown in Fig. 4.7. As can be seen, the acquisition circuit is really simple, as the current error is *directly* acquired by a differential amplifier. This provision, that requires the current reference signal to be *externally* generated, allows to maximize the resolution in the current error internal numerical representation, which, as will be shown, is beneficial for the limitation of limit cycles in the frequency stabilization algorithm. The sampling period, T_{clock}, is N times smaller (with N *even*) than the desired switching period, T_S^*, which defines N as the *oversampling factor* of the hysteresis current controller.

No additional analog hardware is required with respect to what is shown in Fig. 4.7, because both the nonlinear current controller and the frequency regulation strategy are completely "virtualized" and turned into hardware synthesized algorithms. As a result, both a tight-switching frequency regulation and an almost ideal large-signal transient response are possible, and without the drawbacks of a mixed signal implementation. Before describing in detail how the controller operates, we need to briefly analyze the problem of switching frequency stabilization in hysteresis current controllers.

4.3.1   SWITCHING FREQUENCY STABILIZATION

We can illustrate the need for frequency stabilization in hysteresis current controllers referring to Figs. 4.7 and 2.1, neglecting R_S and considering *a single modulation period* of duration $T_S = 1/f_S$. The following current error dynamic equation can be written:

$$\frac{d\,\varepsilon_I(t)}{dt} = \frac{1}{L_S}\left(V_{OC}(t) - E_S(t)\right) - \frac{d\,I_{OREF}(t)}{dt}, \tag{4.9}$$

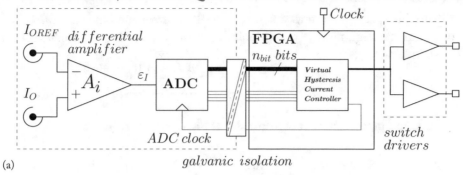

(a)

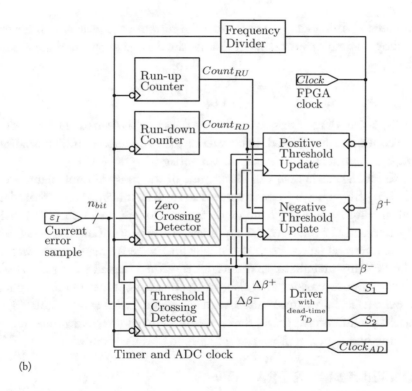

(b)

Figure 4.7: Digital, fixed-frequency hysteresis current controller hardware organization. (a) Current error acquisition circuit and (b) internal FPGA chip organization.

where $t \in [0, T_S]$. Assuming both the current reference I_{OREF} and the output voltage E_S are *slowly* varying during the modulation period, the current error slope expression can be written as:

$$s^{\pm}(t) \cong s^{\pm}(0) = \frac{\pm V_{DC} - E_S(0)}{L_S} - \frac{d\,I_{OREF}(t)}{dt}\bigg|_{t=0}, \qquad (4.10)$$

where all variable quantities are assumed to be well approximated by their *initial* values, taken at $t = 0$, i.e., at the beginning of the modulation period). Defining the instantaneous converter modulation index.

$$h(t) = \frac{E_S(t) + L_S \dfrac{d\, I_{OREF}(t)}{dt}}{V_{DC}}, \tag{4.11}$$

at any modulation period the current error slopes (4.10) can be re-written as:

$$s^{\pm} \cong s^{\pm}(0) = \frac{V_{DC}}{L_S}(\pm 1 - h(0)). \tag{4.12}$$

It is now immediate, from (4.12), to prove that the current error peak amplitude or *threshold*, β^*, that, enforced by a suitable controller, determines the desired duration of the modulation period, T_S^*, is given by:

$$\beta^* = \frac{V_{DC}\, T_S^*}{4\, L_S}\left(1 - h^2(0)\right). \tag{4.13}$$

Equation (4.13) proves that, if constant switching frequency is desired, the current error threshold, β^*, has to be continuously adjusted, to compensate the modulation index variations determined by *non constant* current reference and/or output voltage, as per (4.11).

In the following section we will discuss one of the possible implementations of an *almost constant* switching frequency hysteresis current controller. It is important to underline that, differently from the multi-sampled PI and predictive controllers, the presented material is the result of recent research activity and does not represent in any way a standard, widely applied current control approach. Several aspects of the technique that is used to regulate the switching frequency are still in the experimental phase, so the reader is warned that all what follows is susceptible of modifications and improvements. It is also worth pointing out the peculiar characteristics of the solution, namely: (i) its fully digital nature, with minimum analog components; (ii) its equivalence to a non linear digital filter that drives the converter switches directly from current error samples; and (iii) its being based on a high oversampling ratio of the current error.

4.3.2 CONTROLLER OPERATION

The logic of the control algorithm can be explained referring to Figs. 4.7(b) and 4.8. The former shows the simplified, internal organization of the FPGA circuit that implements the algorithm. The latter, instead, shows the current error signal evolution in a few control cycles around the generic k^{th} iteration of the algorithm, determined by a wrong hysteresis threshold positioning at instant $t_0 = (k-2)T_S^*/2$. To explain, as simply as possible, how the controller brings the current trajectory back to its ideal course, represented by the dashed line, Fig. 4.8 assumes that no switch dead-times, or other data quantization induced errors, affect the circuit operation. As mentioned above, the algorithm's purpose is to mimic an analog hysteresis controller. Besides that, it is designed to synchronize the current error zero crossing instants with a predefined pulse

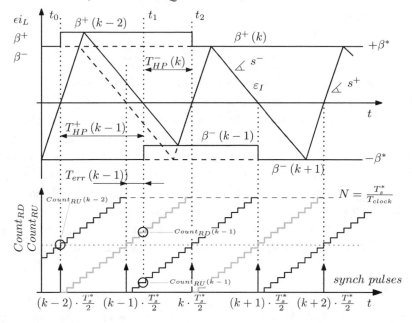

Figure 4.8: Internal variable evolution and control algorithm operation in ideal conditions, i.e., with no errors in threshold application. The sampling clock period is $T_{clock} = T_S^*/N$, where N is the oversampling ratio of the controller. Synchronization pulses occur at each counter's mid-count value, i.e., with $T_S^*/2$ periodicity.

sequence, the *synch pulses* of Fig. 4.8, whose frequency is twice the desired converter switching frequency. To achieve that, at any sampling clock front, i.e., N times in a T_S^* period, a new current error sample is acquired and processed in parallel by the two detection blocks of the FPGA circuit, highlighted in Fig. 4.7(b), to verify if either a zero crossing or a threshold crossing condition has occurred. If neither has taken place, the circuit's state remains unaltered, and it just keeps waiting for a new sample. Otherwise, the appropriate sub-circuits are triggered, as explained in the following.

Zero Crossing Detection

If a zero crossing is detected, one of the two threshold update sub-circuits is triggered, depending on the current error slope. Indeed, a key feature of the proposed control strategy is the use of independently regulated positive and negative hysteresis thresholds. Thanks to the FPGA hardware capabilities, at the time of current error zero crossing, *only the threshold the current error is directed to* is adjusted. As an example, in Fig. 4.8, at instant t_1 the zero crossing detector triggers the *negative* threshold update circuit. The sub-circuit calculates the new threshold level according to the following equation:

$$\beta^-(k-1) = -\frac{T_S^* - 2\,T_{err}(k-1)}{T_S^*}\cdot\beta^*, \tag{4.14}$$

that can be derived immediately from simple geometrical considerations on the similarity of triangles. In (4.14), $T_{err}(k-1)$ represents the *synchronization error* of the last completed switching half-period, i.e., the time distance between the zero crossing instant and the most recent synchronization pulse. In order to measure T_{err}, the threshold update circuit takes advantage of two timers, both with clock period equal to $T_{clock} = T_S^*/N$, associated to the positive (Run Up, RU) and negative (Run Down, RD) slope zero crossings respectively. The timers' setting is such that (i) they are reset when the target modulation period, T_S^*, has elapsed and (ii) there is a half-period, $T_S^*/2$, delay between them.

When the zero crossing condition occurs, both timers are read and the synchronization error is calculated, as an integer number of clock periods, according to the following relation:

$$\frac{T_{err}(k-1)}{T_{clock}} = Count_{RD}(k-1) - \frac{T_S^*}{2\,T_{clock}}, \tag{4.15}$$

where $Count_{RD}(k-1)$ is the reading of the timer at the $(k-1)^{th}$ algorithm iteration, indicated in Fig. 4.8. To finalize the calculation of (4.14), β^*, that is actually *unknown*, has to be determined as well. Because β^* experiences only negligible variations in a $T_S^*/2$ interval, the most recently adjusted value of the opposite current error threshold can be used to determine it, observing that, again for the similarity of triangles,

$$\beta^* = |\beta^+(k-2)|\cdot\frac{T_S^*}{2\,T_{HP}^+(k-1)}, \tag{4.16}$$

where $T_{HP}^+(k-1)$ represents the measured duration of the last completed switching half-period, i.e., the time distance between two consecutive zero crossings of the current error (the superscript indicating that it refers to the positive phase of the current error). To determine $T_{HP}^+(k-1)$, the timers are used once again. Indeed,

$$\frac{T_{HP}^+(k-1)}{T_{clock}} = Count_{RU}(k-1) - Count_{RU}(k-2), \tag{4.17}$$

where $Count_{RU}(\cdot)$ is the reading of the timer at the indicated algorithm iteration. Once $T_{HP}^+(k-1)$ is measured, substituting (4.16) into (4.14) yields the following, final relation:

$$\beta^-(k-1) = -\frac{\frac{T_S^*}{2} - T_{err}(k-1)}{T_{HP}^+(k-1)}\cdot\beta^+(k-2). \tag{4.18}$$

As can be inferred from Fig. 4.8, (4.18) guarantees the next current error zero crossing will be synchronized with the following *synch pulse* at instant $t_2 = kT_S^*/2$. A perfectly symmetrical expression is used by the other sub-circuit to calculate, at the next iteration, the adjusted value of the positive current error threshold, $\beta^+(k)$, based on the run-up phase synchronization error, i.e., $T_{err}(k) \cong 0$, and the half period duration in the negative error phase, i.e., $T_{HP}^-(k)$. When the algorithm reaches the steady state, the current error zero crossings are in phase with the corresponding synchronization pulses and, therefore, the switching period matches the target value, T_S^*. One relevant advantage of (4.18) with respect to other frequency regulation strategies, is that it is completely insensitive to system's parameters (V_{DC} and L_S) possible variations. Indeed, it totally relies on time interval measurements.

Threshold Crossing Detection

The threshold crossing sub-circuit is in charge of checking the relation between the current error sample and the hysteresis thresholds β^+ and β^-. When the current error sample crosses one of them, the appropriate switching action is commanded, so as to determine the current error slope reversal. At the same time, the measurement of the threshold error is initiated. Indeed, in the practical implementation, the current error slope reversal can be delayed with respect to the threshold crossing instant, basically due to the acquisition delay and to inverter dead-times, as shown in the insets of Fig. A8.1. The measurement of the threshold error is then used by the threshold update circuits to correct the threshold level, as explained in Aside 8.

Aside 8. Compensation of Systematic Errors in the Digital Hysteresis Current Controller

The above explained frequency regulation algorithm is exposed to different systematic and random error sources. The main ones are represented by:

1. analog to digital conversion (ADC) delay and dead times;

2. finite counter resolution;

3. threshold saturation;

4. sampling noise

that will be now examined in more detail.

ADC Delay and Dead-Times

As mentioned above, the virtualization of the hysteresis comparator, replaced by a numerical comparison between the digital, quantized representations of current error and hysteresis thresholds, introduces uncertainty in threshold crossing detection and a randomly variable delay in switch commutations, both due to the quantization of the current error. A similar effect is caused by dead-times, that, however, determine a more systematic and much larger error. Altogether, these effects can be indicated as *threshold errors*. Their compensation is mandatory to achieve high quality switching frequency regulation and to keep zero average current error.

To this purpose, when a threshold is crossed, the detector starts calculating and storing the difference between the incoming current error samples and the last crossed threshold level, until it detects a *slope reversal*. The last measured difference represents the threshold error, indicated as $\Delta\beta^{\pm}(\cdot)$ in Fig. A8.1. From this standpoint, the digital, virtual implementation of the hysteresis controller is advantageous with respect to the analog one, where the threshold error due to dead-times has to be determined through interpolation, [6], or approximated estimation, [9]. The calculated error is then passed to the appropriate threshold update sub-circuit, that uses it, at its next activation, to correct the threshold level. As an example, the negative threshold adjustment algorithm, taking into account threshold errors, is modified as:

$$
\begin{aligned}
\beta^{-}(k-1) = -\frac{\frac{T_S^*}{2} - T_{err}(k-1)}{T_{HP}^{+}(k-1)} \cdot \\
\cdot \left[\beta^{+}(k-2) + \Delta\beta^{+}(k-2)\right] - \Delta\beta^{-}(k-3).
\end{aligned}
\tag{A8.1}
$$

A symmetrical expression applies to the positive threshold. Equation (A8.1) is the one actually calculated by the threshold update circuit and is based on the assumption that the threshold error will be approximately invariant in a T_S^* time interval, so that $\Delta\beta^{-}(k-1) \cong \Delta\beta^{-}(k-3)$. Referring once again to Fig. A8.1, it is possible to see the different corrections, with respect to Fig. 4.8 (dashed trace), necessary to compensate for the same initial perturbation of the positive threshold.

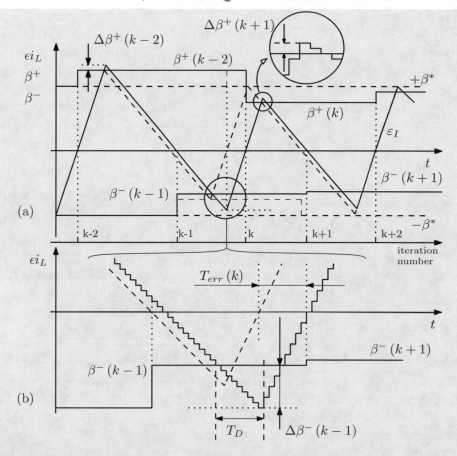

Figure A8.1: (a) Internal variable evolution and control algorithm operation in the presence of dead-times and quantization effects, determining threshold application errors. (b) Detail of (a) where the dead-time induced threshold error is magnified.

After the transient, the negative threshold will be actually set to a less negative value, with respect to the ideal one, $-\beta^*$, fully compensating the dead-time effect on the average current error. This unsymmetrical threshold positioning allows the current error to keep zero average value, which cannot be guaranteed by setting $\beta^+(\cdot) = -\beta^-(\cdot)$. In that case, the regulation would be maintained at the expense of an average current error. Instead, keeping the current error zero crossings synchronized with the reference pulses, zero average current error is guaranteed *together* with frequency regulation.

Finite Counter Resolution

The choice of T_{clock} is crucial to the algorithm operation. Indeed, the algorithm time measurements (4.15) and (4.17) are obtained as integer multiples of this period, that also represents the uncertainty in the measurement of $T_{HP}^{\pm}$. As a result, the algorithm inherently generates limit cycle oscillations, even in ideal conditions (no dead-times, no quantization or threshold errors). In order to limit the LCO amplitude, the ratio between the target switching period, T_S^*, and the clock period must not be set too low. Provided that $\beta^{\pm}(\cdot)$ is represented on more bits than the timing measurements, the minimum oscillation amplitude is easy to predict, using (4.16). It is given by the following relation:

$$\Delta\beta_{LCO} = \pm\frac{2\,T_{clock}}{T_S^*}\cdot\beta^* = \pm\frac{2}{N}\cdot\beta^*, \tag{A8.2}$$

that can be used as a basic guideline to choose the maximum applicable T_{clock}. Because β^* and the switching period are proportional to each other, (A8.2) proves that the best case cycle by cycle relative error on frequency regulation is $\pm 2/N$. In general, a certain amplification of the LCO amplitude can take place, especially when the regulation algorithm operates close to the maximum modulation index, because of other non-linear effects (e.g., saturations) or threshold errors. Therefore, (A8.2) gives just a best case estimation of the steady state frequency regulation error.

On the other hand, increasing the sampling rate and clock resolution beyond a maximum limit, that depends on the slope (4.12) of the current error and on the resolution of its numerical representation, results in multiple samples falling into the same ADC bin. When this happens, there is no real advantage in further increasing the clock frequency; the uncertainty in zero or threshold crossing detections will become higher than T_{clock}. A criterion to estimate the upper limit of the controller's clock frequency is given by:

$$f_{clock}^{\max} = \frac{1}{T_{clock}^{\min}} = \frac{V_{DC}}{L_s}\cdot\frac{2^{n_{bit}+1}}{I_{MAX}}, \tag{A8.3}$$

with the usual definitions of the parameters. The meaning of (A8.3) is that, when the modulation index is zero, the current error slope will be such that one LSB will be added for each clock period to the current error representation. As a result, zero or threshold crossing will be detected with a single clock period uncertainty.

Threshold Saturation

When the converter operates at high modulation index, the conduction interval of either S_1 or S_2 becomes relatively short. Because the algorithm takes a definite amount of time, which we denote as latency time T_{lat}, to compute and adjust the current error threshold after each zero crossing of the current error, if the run-up (or the run-down) phase becomes too short, positive (or negative) synchronization errors cannot be corrected and undesired transients are generated. In addition, if the run-up phase, or the run-down phase, becomes shorter than twice the dead-time T_D, the compensation is altogether impossible, as thresholds cannot invert their sign. Therefore, the algorithm can operate as described above only up to a maximum modulation index level, which is given by:

$$M_{\max} = \min\left\{1 - 4\,\frac{T_D}{T_S^*},\ 1 - 4\,\frac{T_{lat}}{T_S^*}\right\}, \tag{A8.4}$$

where T_{lat} is equal to 550 ns in our hardware, determining a 0.956 maximum modulation index. To prevent undesired oscillations of thresholds when the modulation index tends to become higher than (A8.4), a saturation strategy needs to be implemented. In our case, any time the calculated threshold becomes lower than a predefined minimum, fixed threshold regime is entered, i.e., the last non saturated threshold value is applied until the modulation index reduces and threshold calculation no longer results into too small values. Of course, frequency regulation is lost during saturation, but, thanks to the unsymmetrical threshold positioning, average current control can be maintained.

Sampling Noise

Because the current controller operates comparing current samples to thresholds and measuring zero crossing synchronization errors, a good signal to noise ratio (SNR), in the current error sampling process is mandatory to achieve satisfactory performance levels. The algorithm can be made somewhat tolerant to input injected noise by introducing hysteretic nonlinearities in the zero crossing and threshold crossing detection sub-circuits and, sacrificing some bandwidth, by implementing an input digital denoising filter.

Nevertheless, if the noise level is too high, the resulting jitter in the zero crossing and threshold crossing detections will rapidly deteriorate the performance. To estimate the controller robustness, simulations have been performed injecting white noise of increasing power at the ADC input. Doing so, it was possible to verify that, as long as the ADC output SNR is above 60 dB, i.e., the controller operates on at least 10 effective bits in the current error representation, excellent frequency stabilization capabilities (average error below 1%, with an adequate value of N) can be obtained. To guarantee this SNR level, care must be taken in the acquisition board design and layout. Analog filters can be used as well, but their frequency response needs to be carefully shaped, so as to avoid significant current error waveform distortion.

The hysteresis controller design for the test bench converter led to the parameters listed in Table 4.4. It is worth noting that the oversampling factor was increased to $N_{Hyst} = 100$; this is necessary to guarantee a sufficiently tight control of the converter switching frequency, that, according to (A8.2), will be kept close to its set-point, $f_S^* = 20$ kHz, with an expected $\pm 2\%$ accuracy.

Table 4.4: Hystereis current controller parameters

Parameter	Symbol	Value	
Clock frequency	f_{clock}	2	MHz
Target switching frequency	f_s^*	20	kHz
Oversampling rate	N_{Hyst}	100	
AD converter latency	Δt_{AD}	0.2	μs
Current sensor gain	G_{TI}	100	mV/A
Circuit latency (worst case)	T_{lat}	550	ns

4.3.3   SMALL-SIGNAL FREQUENCY RESPONSE

As mentioned above, because of the non linear nature of the hysteresis current controller, it is not possible to derive any small-signal transfer function analytically. However, it is possible to set-up a detailed numerical simulation of the controller operation and to emulate the small-signal response test. For better accuracy, all the implementation details, including quantizations, rounding effects and finite arithmetic resolution have to be included in the simulation model. Once the simulation results are collected, they can be compared to the experimental ones. The outcome is presented in Fig. 4.9. It is interesting to observe how experiments match quite accurately the numerical simulation data. From both it seems evident that the hysteresis current controller offers a flat response to small perturbations of the reference signal.

This interesting result can be explained considering that the objective of the non linear controller, as explained in Section 4.3, is to keep the current error zero crossings synchronized with a reference clock, compensating any perturbation in a single half modulation period. As a consequence, any perturbation of the steady state is rejected in the current modulation half period, which implies virtually no small-signal delay in the frequency span of interest. Indeed, the phase shift is minimum for this controller, being equal to $-3°$ at 3.0 kHz. At the same time, the magnitude response is almost flat.

4.4   LARGE-SIGNAL RESPONSE TEST

We would like now to provide some illustration of the three considered controllers' large-signal response performance. As we will see in Chapter 6, when the current controller is used to support an outer voltage control loop, its capability to track fast reference signal variations is of paramount importance in determining the final performance of the voltage loop.

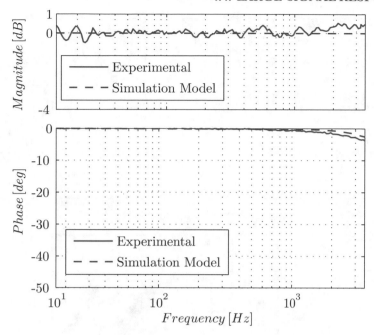

Figure 4.9: Digital, fixed frequency hysteresis current controller small-signal response in magnitude (upper plot) and phase (lower plot) according to: numerical simulations (dashed) and experiments (solid).

Therefore, it is interesting to evaluate the large-signal responses of the three considered multi-sampled controllers. To this purpose, two different tests have been performed: (i) the step variation of a DC nominal current reference from 0 A to 10 A and from 10 A to 0 A; and (ii) a sequence of ±180° phase step variations in a sinusoidal reference signal.

In the DC current step variation experiments, the converter output voltage E_S has been externally controlled to a constant DC level, so as to achieve constant 0.8 modulation index, defined as per (4.11), operation. Similarly, in the AC step response tests, the inverter has been connected to an AC voltage source, imposing $E_S = 140$ V RMS, i.e., once again, a 0.8 peak modulation index.

The obtained experimental results, for the oversampled PI controller, are shown in Fig. 4.10, while those referring to the oversampled predictive controller are shown in Fig. 4.11. Finally, the hysteresis controller performance is illustrated in Fig. 4.12.

In the fastest transients, shown in Figs. 4.10(b), 4.11(b), and 4.12(b) we can see that the hysteretic and the predictive controllers offer the best responsiveness, while the worst is the one given by the PI controller. Thanks to the oversampled acquisition of the current signal, in all cases, the reference step change is detected with minimum delay, i.e., in one sampling period.

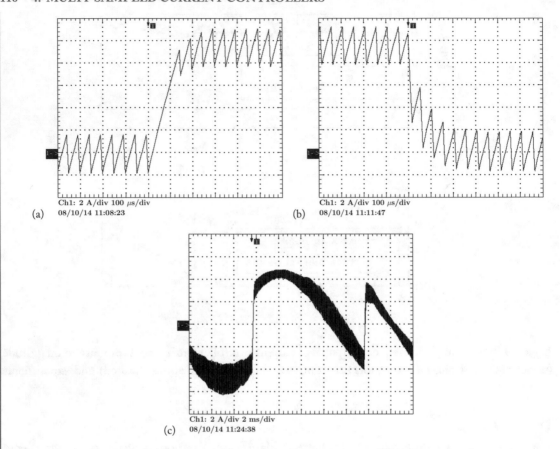

(a)

(b)

(c)

Figure 4.10: Experimental measurement of the PI controller large-signal step responses: (a) response to a +10 A step of the reference current; (b) response to a −10 A step of the reference current; (c) response to 180° phase jumps in a sinusoidal reference current.

Consistent results are obtained, as well, in a condition equivalent to a grid-tied operation of the VSI, illustrated by Figs. 4.10(c), 4.11(c) and 4.12(c).

It is certainly worth comparing these results with those discussed in previous chapters. For example, if we compare Fig. 4.10(b) with Fig. 2.11(d), it is possible to appreciate how the multi-sampled PI behaves very similarly to an analog PI. This means that the digital PWM and computational delay bottlenecks have both been effectively overcome by the multi-sampled organization.

Instead, comparing Fig. 4.11(b) with Fig. 3.18(d), it is possible to see how the multi-sampled dead-beat is capable of faster reaction times and equally clean transient responses. In this case, the multi-sampled implementation allows the predictive control algorithm to adjust the

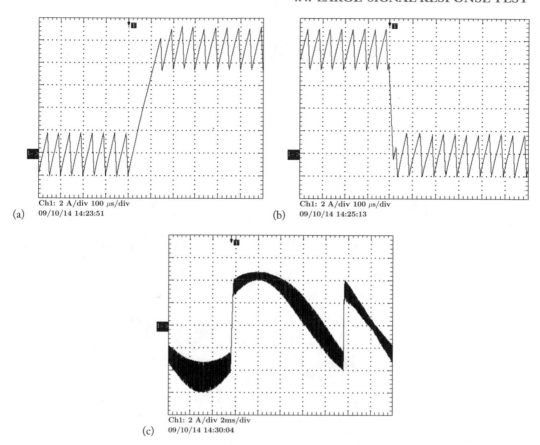

Ch1: 2 A/div 100 μs/div
(a)   09/10/14 14:23:51

Ch1: 2 A/div 100 μs/div
(b)   09/10/14 14:25:13

Ch1: 2 A/div 2ms/div
(c)   09/10/14 14:30:04

Figure 4.11: Experimental measurement of the predictive controller large-signal step responses: (a) response to a +10 A step of the reference current; (b) response to a −10 A step of the reference current; and (c) response to 180° phase jumps in a sinusoidal reference current.

duty-cycle instantaneously, so that it is not necessary to wait for a full modulation period to see the controller respond to the reference variation.

A similar outcome is determined for the virtual hysteresis controller, which is capable of very low delay responses. Some relatively longer transient is determined by the frequency adjustment algorithm, that takes a few modulation periods to lock to synchronization pulses after the transient. But this transient, it is worth recalling, does not affect the average current regulation, that is resumed as soon as the inverter starts switching again.

A final experimental test, that can be of some interest for grid tied VSI applications, consists in evaluating the controllers' reference tracking capability with respect to a low-frequency (e.g., 50 Hz) sinusoidal reference signal and measuring the resulting, steady-state total harmonic

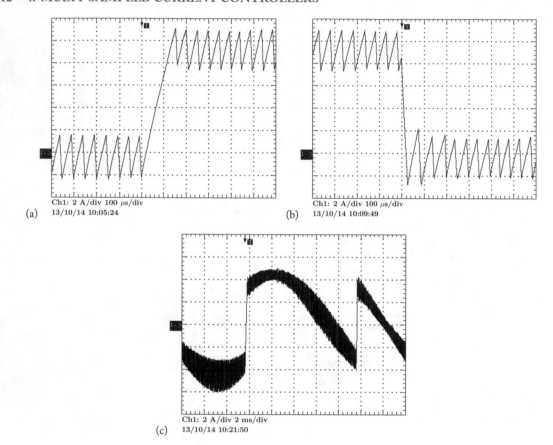

Figure 4.12: Experimental measurement of the hysteresis controller large-signal step responses: (a) response to a $+10$ A step of the reference current; (b) response to a -10 A step of the reference current; and (c) response to 180° phase jumps in a sinusoidal reference current.

distortion (THD). The test is performed applying a distorted sinusoidal voltage (with 2% third harmonic and 1% fifth and seventh harmonic components) as the phase output voltage E_S and controlling the inverter so as to obtain a 10 A peak amplitude sinusoidal current, in phase with the fundamental component of E_S. The output current THD has been calculated considering harmonic components up to order 40. The outcome is reported in the last row of Table 4.5, that summarizes the most relevant results of all the performed experiments.

It is possible to notice how all the controllers present very low distortion levels, significantly lower than the E_S voltage's ($\cong 2.5\%$). Comparing the values, we can also see that the predictive controller performs better than the others and that the PI controller is the one that obtains the highest distortion. That is because the wider small-signal bandwidth of the predictive and the

Table 4.5: Summary of the experimental results

	PI	Predective	Hysteretic
Phase shift @ 0.5 kHz	$-8°$	$-5°$	$-0°$
Phase shift @ 1.0 kHz	$-18°$	$-10°$	$-1°$
Phase shift @ 2.0 kHz	$-37°$	$-19°$	$-2°$
Phase shift @ 3.0 kHz	$-49°$	$-28°$	$-3°$
THD of 50Hz sinusoid	1.52%	0.82%	1.47%

hysteretic controllers allows them to better reject the E_S voltage harmonics with respect to the PI controller. On the other hand, the instantaneous switching frequency control of the hysteresis controller is itself responsible for a slight increase of the inverter current harmonic content (due to ripple amplitude modulation), which reflects into a worse performance as compared to the predictive controller's.

4.5   FPGA CHIP UTILIZATION

To conclude our brief discussion of the multi-sampled controllers, it is quite interesting to compare their computational burden, and the consequent FPGA chip utilization. As mentioned above, the PI and predictive controller are very simple to implement and require the minimum amount of hardware resources, less than 5% of both DSP units and logic slices (combinatorial and sequential logic blocks) of the adopted Spartan 6 LX 45 chip. For the same reason, they also offer minimum computational delays, equal to one FPGA clock period, that could be as low as 5 ns, but is actually set to 50 ns in our implementation.

Instead, the hysteresis controller is relatively more complex, essentially because of the frequency regulation algorithm and the input and current error de-noising filters. For this reason, the FPGA clock period must be increased or multiple periods must be allocated to complete the threshold regulation algorithm. Overall, a higher worst-case latency has been measured, in the order of 0.55 μs corresponding to 11 FPGA clock cycles at the considered clock frequency (20 MHz). This is mainly due to the calculation of the divisions required by the frequency regulation algorithm (A8.1). As far as FPGA chip utilization is concerned, less than 7% of both DSP units and logic slices were used. Overall, the three solutions are compatible with tiny FPGA chips.

4.6   MULTI-SAMPLED CURRENT CONTROLLERS: CONCLUSIONS

The analytical and experimental results presented in this chapter show that, once a non conventional control hardware comes into play, new interesting solutions become viable, whose performance can be significantly higher than the conventional ones.

The current controller of a VSI, accordingly, can achieve higher small-signal bandwidth and large-signal speed of response, which can be beneficial to the design of possible outer control loops. In some very demanding applications, this can even be the only acceptable solution.

On the other hand, the implementation is no longer as simple as in the case of software controllers and there is no standard, ready to use hardware control platform to rely on. Quite on the contrary, custom control platforms comprising, at least, a FPGA chip and a fast ADC have to be developed, which can be a time consuming and relatively complex additional phase of the controller design process. As we will see in Chapter 7, however, this is no longer the only viable option as, more and more often, complex multi-platform control devices can be found on the market, at reasonable prices, that allow to rapidly develop FPGA applications.

More than ever, it is important to carefully evaluate the actual hardware requirements, determined by each application specifications, so as to implement the most appropriate solution, correctly balancing cost and performance. A good design should indeed comply with the specifications at the minimum cost, i.e., without using more than the strictly required resources.

REFERENCES

[1] S. Buso, T. Caldognetto, D.I. Brandao: "Oversampled Deadbeat Current Controller for Voltage Source Converters," IEEE Applied Power Electronics Conference (APEC), Charlotte, NC, USA, 2015, pp. 1493–1500. 86, 96

[2] S. Buso, T. Caldognetto: "A Non-linear Wide Bandwidth Digital Current Controller for DC-DC and DC-AC Converters," IEEE Industrial Electronics Conference (IECON), Dallas, TX, USA, 2014, pp. 1090–1096. DOI: 10.1109/IECON.2014.7048638. 86, 98

[3] A.V. Peterchev, S.R. Sanders, "Quantization resolution and limit cycling in digitally controlled PWM converters," *IEEE Transactions on Power Electronics*, Vol. 18, No. 1, January 2003, pp. 301–308. DOI: 10.1109/TPEL.2002.807092. 89

[4] M. Bradley, E. Alarcon, O. Feely, "Design-Oriented Analysis of Quantization - Induced Limit Cycles in a Multiple-Sampled Digitally Controlled Buck Converter," *IEEE Transactions on Circuits and Systems I*, Regular Papers, Vol. 61, No. 4, April 2014, pp. 1192–1205. DOI: 10.1109/TCSI.2013.2283671. 89

[5] L. Corradini, P. Mattavelli, "Modeling of Multisampled Pulse Width Modulators for Digitally Controlled DC-DC Converters," *IEEE Transactions on Power Electronics*, Vol. 23, No. 4, July 2008, pp. 1839–1847. DOI: 10.1109/TPEL.2008.925422. 88

[6] S. Buso, S. Fasolo, L. Malesani, P. Mattavelli: "A Dead-Beat Adaptive Hysteresis Current Control," *IEEE Transactions on Industry Applications*, Vol. 36, No. 4, July/August 2000, pp. 1174–1180. DOI: 10.1109/28.855976. 104

[7] C.N.M. Ho, V.S.P. Cheung, H.S.H. Chung, "Constant-Frequency Hysteresis Current Control of Grid-Connected VSI Without Bandwidth Control," *IEEE Transactions on Power Electronics,* Vol. 24, No. 11, November 2009, pp. 2484–2495. DOI: 10.1109/TPEL.2009.2031804.

[8] J.C. Olivier, J.C. Le Claire and L. Loron, "An Efficient Switching Frequency Limitation Process Applied to a High Dynamic Voltage Supply," *IEEE Transactions on Power Electronics,* Vol. 23, No. 1, January 2008, pp. 153–162. DOI: 10.1109/TPEL.2007.911876.

[9] D.G. Holmes, R. Davoodnezhad, B.P. McGrath, "An Improved Three-Phase Variable-Band Hysteresis Current Regulator," *IEEE Transactions on Power Electronics,* Vol. 28, No. 1, January 2013, pp. 441–450. DOI: 10.1109/TPEL.2012.2199133. 104

[10] L. Fangrui, A.I. Maswood, "A Novel Variable Hysteresis Band Current Control of Three-Phase Three-Level Unity PF Rectifier With Constant Switching Frequency," *IEEE Transactions on Power Electronics,* Vol. 21, No. 6, November 2006, pp. 1727–1734. DOI: 10.1109/TPEL.2006.882926.

CHAPTER 5

Extension to Three-Phase Inverters

In this chapter we present the possible means for the application to three phase inverters of what we have just seen about digital current control of single-phase VSIs. When the three phase converter is characterized by four wires, i.e., three phases plus neutral, the application is straightforward, since a four wire three phase system is totally equivalent to three independent single phase systems. Of course, this particular situation does not require any further discussion. On the contrary, we need to apply a little more caution when we are dealing with a three phase system with insulated neutral, i.e., with a three-wire, three-phase system. The objective of this chapter to give the basic knowledge needed to extend the control principles we have previously described to this kind of systems. Two fundamental tools are required to design an efficient three phase current controller: (i) $\alpha\beta$ transformation and (ii) space vector modulation (SVM).

In the first part of this chapter, we are going to illustrate the principles of both. Successively, we will show how, under certain assumptions, the three-phase system dynamic model can be transformed into an equivalent two-phase system, with independent, i.e., orthogonal, components. We will see how, in this particular case, the controller design for the two phase system is identical to that of a single phase one.

In the final part of this chapter we will discuss a particular kind of two phase controller, known as *rotating reference frame* controller, presenting the merits and limitations of this solution.

5.1 THE $\alpha\beta$ TRANSFORMATION

The $\alpha\beta$ transformation represents a very useful tool for the analysis and the modeling of three-phase electrical systems. In general, a three-phase linear electric system can be properly described in mathematical terms only by writing a set of tri-dimensional dynamic equations (integral and/or differential), providing a self consistent mathematical model for each phase. In some cases, though, the existence of physical constraints makes the three models not independent from each other. In these circumstances the order of the mathematical model can be reduced without any loss of information. We will see a remarkable example of this in the following.

Supposing that it is physically meaningful to reduce the order of the mathematical model from three to two dimensions, $\alpha\beta$ transformation represents the most commonly used relation to perform such order reduction. To explain the way it works we can consider a tri-dimensional vector $\vec{x}_{abc} = \begin{bmatrix} x_a & x_b & x_c \end{bmatrix}^T$ that can represent any triplet of system's electrical variables (voltages or currents). We can now consider the following linear transformation, $T_{\alpha\beta\gamma}$,

$$\begin{bmatrix} x_\alpha \\ x_\beta \\ x_\gamma \end{bmatrix} = T_{\alpha\beta\gamma} \begin{bmatrix} x_a \\ x_b \\ x_c \end{bmatrix} = \sqrt{\frac{2}{3}} \begin{bmatrix} 1 & -1/2 & -1/2 \\ 0 & \sqrt{3}/2 & -\sqrt{3}/2 \\ 1/\sqrt{2} & 1/\sqrt{2} & 1/\sqrt{2} \end{bmatrix} \begin{bmatrix} x_a \\ x_b \\ x_c \end{bmatrix} \quad (5.1)$$

that, in geometrical terms, represents a change from the set of reference axes denoted as abc to the equivalent one indicated as $\alpha\beta\gamma$. This change of reference axes takes place because the standard $\mathbb{R}^3$ orthonormal base B_{abc},

$$B_{abc} = \left\{ \begin{bmatrix} 1 & 0 & 0 \end{bmatrix}^T, \begin{bmatrix} 0 & 1 & 0 \end{bmatrix}^T, \begin{bmatrix} 0 & 0 & 1 \end{bmatrix}^T \right\}, \quad (5.2)$$

is replaced by the new base $B_{\alpha\beta\gamma}$,

$$B_{\alpha\beta\gamma} = \sqrt{2/3} \left\{ \begin{bmatrix} 1 & -1/2 & -1/2 \end{bmatrix}^T, \right.$$
$$\left. \begin{bmatrix} 0 & \sqrt{3}/2 & -\sqrt{3}/2 \end{bmatrix}^T, \begin{bmatrix} 1/\sqrt{2} & 1/\sqrt{2} & 1/\sqrt{2} \end{bmatrix}^T \right\}. \quad (5.3)$$

The $B_{\alpha\beta\gamma}$ base is once again orthonormal, i.e., its vectors have unity norm and are orthogonal to one another, thanks to the presence of the $\sqrt{2/3}$ coefficient, that also appears in (5.1). This coefficient is sometimes omitted when keeping the base orthonormality is not considered essential. However, orthonormality implies that: (i) the inverse of the $T_{\alpha\beta\gamma}$ transformation is equal to the matrix transposed, i.e., $T_{\alpha\beta\gamma}^{-1} = T_{\alpha\beta\gamma}^T$, and (ii) the computation of electrical powers is independent from the transformation of coordinates, i.e., $\langle \vec{e}_{abc}, \vec{i}_{abc} \rangle = \langle \vec{e}_{\alpha\beta\gamma}, \vec{i}_{\alpha\beta\gamma} \rangle$, where the "$\langle \rangle$"operator represents the scalar product of vectors, $\vec{e}$ is a voltage vector and $\vec{i}$ is a current vector. The latter property justifies the fact the (5.1) is sometimes called "power invariant" transformation. The geometrical interpretation of (5.1) is shown in Fig. 5.1(a).

The $T_{\alpha\beta\gamma}$ transformation has an additional, interesting property, that becomes clear when we take into account the following condition

$$x_a + x_b + x_c = 0 \quad \Rightarrow \quad x_\gamma = 0, \quad (5.4)$$

whose meaning is to operate the restriction of the tri-dimensional space to a plane π (Fig. 5.1(a)). Examining (5.3) and (5.4) we can see how the first two components of base $B_{\alpha\beta\gamma}$ lay on π, while the third is orthogonal to π. This means that the first two components of $B_{\alpha\beta\gamma}$ actually represent an orthonormal base for plane π, while the third component has no projection on π. This observation is fundamental for our conclusion: every time the constraint (5.4) is meaningful for a tri-dimensional system, the coordinate transformation $T_{\alpha\beta\gamma}$ allows to describe the same system in a bi-dimensional space without any loss of information. This holds because any vector complying

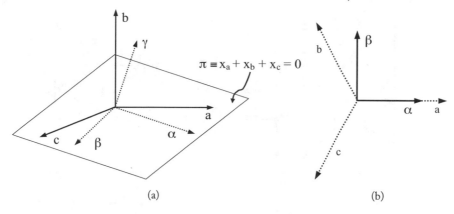

Figure 5.1: (a) Graphical representation of the $T_{\alpha\beta\gamma}$ coordinate transformation.

with (5.4) is actually laying on plane π, and, as such, can be expressed as a linear combination of base vectors defined for π. We can therefore define the so called $\alpha\beta$ transformation as follows:

$$
\begin{bmatrix} x_\alpha \\ x_\beta \end{bmatrix} = T_{\alpha\beta} \begin{bmatrix} x_a \\ x_b \\ x_c \end{bmatrix} = \sqrt{\frac{2}{3}} \begin{bmatrix} 1 & -1/2 & -1/2 \\ 0 & \sqrt{3}/2 & -\sqrt{3}/2 \end{bmatrix} \begin{bmatrix} x_a \\ x_b \\ x_c \end{bmatrix}, \tag{5.5}
$$

and its inverse as

$$
\begin{bmatrix} x_a \\ x_b \\ x_c \end{bmatrix} = T_{\alpha\beta\gamma}^T \begin{bmatrix} x_\alpha \\ x_\beta \\ 0 \end{bmatrix} = \sqrt{\frac{2}{3}} \begin{bmatrix} 1 & 0 \\ -1/2 & \sqrt{3}/2 \\ -1/2 & -\sqrt{3}/2 \end{bmatrix} \begin{bmatrix} x_\alpha \\ x_\beta \end{bmatrix} = T_{\alpha\beta}^T \begin{bmatrix} x_\alpha \\ x_\beta \end{bmatrix}. \tag{5.6}
$$

Equation (5.6) is easily obtained exploiting the base orthonormality and considering the transposed matrix of (5.5). In geometrical terms (5.5) simply determines the projection of any vector $\vec{x}_{abc} = \begin{bmatrix} x_a & x_b & x_c \end{bmatrix}^T$ on plane π. We need to underline once more that this is physically meaningful only if the γ component of vector $\vec{x}_{abc} = \begin{bmatrix} x_a & x_b & x_c \end{bmatrix}^T$ is zero. The γ component, as can be easily verified, is nothing but the *arithmetic average* of the three vector component values, also known as *common mode* vector component. When this is not zero, the application of $\alpha\beta$ transformation implies the loss of the information associated to the common mode. It is also interesting to note that the projection on plane π of base B_{abc} determines three 120° angled axes, as shown in Fig. 5.1(b), which makes the matrix in (5.5) easy to remember.

It is very useful to visualize the effect of the application of $T_{\alpha\beta}$ to some particular cases. We begin presenting the case of sinusoidal voltage signals. If we consider a triplet of symmetric

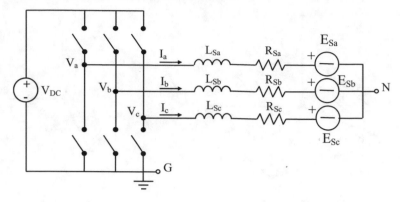

Figure 5.2: Three-phase VSI simplified schematic.

sinusoidal signals like:

$$e_a = U_M \sin(\omega t)$$
$$e_b = U_M \sin(\omega t - 2\pi/3)$$
$$e_c = U_M \sin(\omega t + 2\pi/3)$$

(5.7)

it is easy to verify that

$$e_\alpha = \sqrt{\frac{3}{2}} \, U_M \sin(\omega t)$$
$$e_\beta = -\sqrt{\frac{3}{2}} \, U_M \cos(\omega t).$$

(5.8)

It is possible to see that the space vector $\vec{e}_{abc}$, associated to (5.7), satisfies constraint (5.4) and that, as such, it can be described without loss of information in the $\alpha\beta$ reference frame. In that frame, the vector $\vec{e}_{\alpha\beta}$, can be interpreted as a $\sqrt{\frac{3}{2}} \, U_M$ amplitude rotating vector, the angular rotation speed being equal to ω.

5.2   SPACE VECTOR MODULATION

Space Vector Modulation (SVM) is a frequently used method to implement PWM in three phase switching converters with insulated neutral. It allows not only to simplify the control organization, but also to maximize the exploitation of the converter hardware, inherently realizing a third harmonic injection mechanism. The basic principles behind SVM can be explained referring to the idealized three phase voltage source inverter of Fig. 5.2. As can be seen, the structure is a straightforward extension of the single-phase one we have been considering so far. Its characteristics and modes of operation are analyzed in detail in every power electronics text book (like [1] and [2]), so we won't spend many words on it. However, three fundamental characteristics, essential to the understanding of what follows, have to be underlined: (i) the converter has insulated neutral, i.e., the circuit node indicated by N in Fig. 5.2 is floating, (ii) there is a single input DC voltage source, which makes the phase voltages V_a, V_b, V_c, referred to node G, unipolar, and (iii) the load is generally symmetrical and balanced, i.e., all impedances have the same values and E_{Sa}, E_{Sb} and E_{Sc} are symmetrical and balanced sinusoidal voltages.

The application of SVM requires the instantaneous inverter output voltage, represented by vector $\vec{V}_{abc} = \begin{bmatrix} V_a & V_b & V_c \end{bmatrix}^T$ to be *projected* on the $\alpha\beta$ reference frame, as defined in the previous section. From Fig. 5.2 it is immediately recognizable that, at any instant, each inverter phase voltage can be either zero or equal to the DC link voltage V_{DC}. Therefore, the inverter output voltage vector can assume, at any instant, only one out eight different values. The possible output voltage vector values and their *projections* on plane π are shown in Fig. 5.3. As can be seen, there are two different possibilities to impose a zero phase to phase voltage on the load. This property can be exploited in the implementation of SVM, for example, to minimize the number of switch commutations.

The idea behind SVM is quite simple [3, 4]. A desired output voltage vector, represented in the $\alpha\beta$ reference frame, is obtained from the superposition of the inverter output vectors, so that, on average, at the end of any modulation period a voltage equal to the desired one will have been generated. The procedure can be explained referring to Fig. 5.4. The desired vector, $\vec{V}_{\alpha\beta}^*$, is projected on the two closest inverter output state vectors, i.e., $\vec{V}_{100}$ and $\vec{V}_{110}$, in the example of Fig 5.4. Of course, the position of $\vec{V}_{\alpha\beta}^*$ considered in this example is arbitrary; however, exactly the same reasoning can be applied to different vector locations. The length of each projection, V_1 and V_2, determines the fraction δ of the modulation period that will be occupied by each output vector, according to the following relation:

$$\delta_1 = \frac{|V_1|}{\left|\vec{V}_{100}\right|} \qquad \delta_2 = \frac{|V_2|}{\left|\vec{V}_{110}\right|}. \tag{5.9}$$

The application of the zero voltage vector for a fraction δ_3 of the modulation period is normally required to satisfy the following condition:

$$\delta_1 + \delta_2 + \delta_3 = 1, \tag{5.10}$$

that simply expresses the fact that the modulation period must be fully occupied by output voltage vectors. Following this procedure, the average inverter output voltage $\overline{V}_O$ will be given by

$$\overline{V}_O = \delta_1 V_{100} + \delta_2 V_{110} + \delta_3 V_{111} = V_1 + V_2 = \vec{V}_{\alpha\beta}^*, \tag{5.11}$$

as expected. Please note that: (i) the zero vector can be either $\vec{V}_{111}$ or, equivalently, $\vec{V}_{000}$, (ii) the order of application of the inverter output vectors is arbitrary and can be used as a degree of

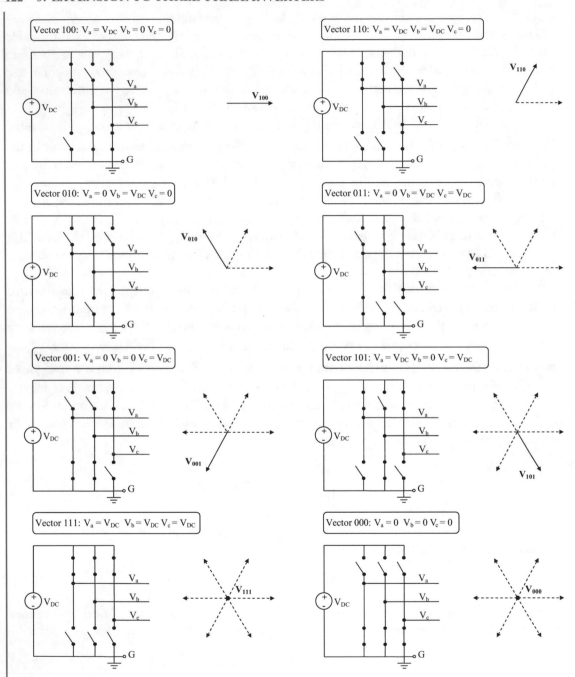

Figure 5.3: Three-phase inverter output voltage vectors and their projection on plane π.

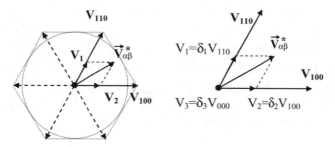

Figure 5.4: Generation of the voltage reference vector by superposition of inverter output vectors.

freedom in the implementation of SVM (see Aside 9), and (iii) the commutation from V_1 to V_2 always requires the commutation of a single inverter phase, no matter what sector of the hexagon the reference vector is laying on.

The implementation of the above-described procedure requires a not negligible amount of computations. In any modulation period, given the α and β components of the voltage reference vector $\vec{V}_{\alpha\beta}^*$ one has to: (i) locate the two closest inverter output vectors, i.e., the hexagon sector where $\vec{V}_{\alpha\beta}^*$ is laying on, (ii) determine the amplitude of V_1 and V_2, and (iii) calculate the values of δ_1, δ_2, δ_3, using (5.9) and (5.10). Of course, the simplest way to perform these computations is by using a microcontroller or DSP. This is the reason why SVM is almost always associated to digital control. In the Aside 9, we will further discuss some implementation issues of SVM.

We have seen in the previous section that the projection on the $\alpha\beta$ reference frame of a triplet of symmetrical, sinusoidal, phase voltages is a constant amplitude rotating vector. Therefore, every time our three phase VSI has to generate a triplet of sinusoidal phase voltages, which happens very frequently, the SVM procedure will have to synthesize the rotating reference vector corresponding to it. This will determine a period by period adjustment of the output vectors and of the δ_1, δ_2, δ_3 values. It can be interesting to identify the locus of the constant amplitude rotating reference vectors that can be generated by the inverter without distortion. This is represented by the circle *inscribed* in the vector hexagon (Fig. 5.4). It is easy to verify that every vector that lies inside the circle generates a valid δ_1, δ_2, δ_3, triplet. Instead, a vector that lies partially outside the circle cannot be generated by the inverter, because the sum of the corresponding δ_1, δ_2, δ_3 becomes greater than unity. This situation is called inverter *saturation* and generally causes output voltage distortion.

If we consider (5.5) and Fig. 5.4, it is easy to calculate the amplitude U_{MMAX} of the voltage triplet (5.7) that corresponds to a rotating vector having an amplitude equal to the radius of the inscribed circle. We find:

$$\sqrt{\frac{3}{2}}U_{MMAX} = \sqrt{\frac{2}{3}}V_{DC}\frac{\sqrt{3}}{2} \quad \Leftrightarrow \quad U_{MMAX} = \frac{2}{\sqrt{3}}\frac{V_{DC}}{2} \cong 1.15 \cdot \frac{V_{DC}}{2}, \tag{5.12}$$

that shows a very interesting fact: the application of SVM increases the range of the possible sinusoidal output voltages by 15% with respect to what could be expected looking at the schematic of Fig. 5.2. The reason why this happens is the following: the inverter output voltage vectors of Fig. 5.3 *do not* comply with constraint (5.4), as it is easy to verify. Consequently, what is used to synthesize the desired output voltage vector $\vec{V}_{\alpha\beta}^{*}$ is not the superposition of vectors laying on plane π, as Fig. 5.4 might suggest. A more realistic representation of the inverter output vectors, that puts into evidence their γ component, is shown in Fig. 5.5.

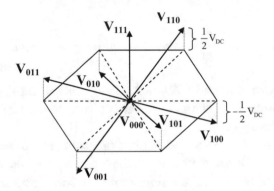

Figure 5.5: Tri-dimensional view of the space vector hexagon.

Aside 9. Implementation of Space Vector Modulation

We now consider a possible implementation algorithm for space vector modulation, which can be directly programmed into a microcontroller or a digital signal processor. The first issue in SVM implementation is the identification of the hexagon sector where the reference vector is lying. This can be done by implementing once again a base change from the $\alpha\beta$ reference frame to a new set of three different reference frames. Figure A9.1 shows the considered set.

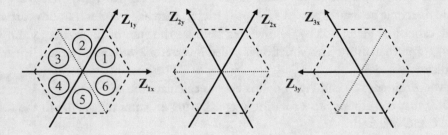

Figure A9.1: Set of three bi-dimensional reference frames.

As can be seen, each frame refers to a particular couple of hexagon sectors. The method we propose simply requires the projection of the inverter output voltage reference vector $\vec{V}_{\alpha\beta}^*$ onto each one of the three hexagon reference frames. This is easily implemented with the following set of reference base change matrices:

$$
M_1 = \begin{bmatrix} 1 & -\dfrac{1}{\sqrt{3}} \\ 0 & \dfrac{2}{\sqrt{3}} \end{bmatrix} \qquad M_2 = \begin{bmatrix} 1 & \dfrac{1}{\sqrt{3}} \\ -1 & \dfrac{1}{\sqrt{3}} \end{bmatrix} \qquad M_3 = \begin{bmatrix} 0 & \dfrac{2}{\sqrt{3}} \\ -1 & -\dfrac{1}{\sqrt{3}} \end{bmatrix}, \qquad \text{(A9.1)}
$$

that map the orthogonal set of axes α and β onto the three, non orthogonal sets Z_i, $i \in \{1, 2, 3\}$. It is interesting to note that the algorithm required to implement the three projections is quite simple. Here we propose a possible operation sequence that gives the six $Z_{ix}Z_{iy}$ components:

$$tmp = \frac{V_\beta^*}{\sqrt{3}}; \qquad \text{save to a temporary register}$$

$$Z_{1x} = V_\alpha^* - tmp; \qquad Z_{1x} \text{ found}$$

$$Z_{2y} = -Z_{1x}; \qquad Z_{2y} \text{ found}$$

$$Z_{1y} = 2tmp; \qquad Z_{1y} \text{ found}$$

$$Z_{3x} = Z_{1y}; \qquad Z_{3x} \text{ found}$$

$$Z_{2x} = V_\alpha^* + tmp; \qquad Z_{2x} \text{ found}$$

$$Z_{3y} = -Z_{2x}; \qquad Z_{3y} \text{ found}$$

As can be seen, the sequence implies the execution of only one multiplication. Once the $Z_{ix}Z_{iy}$ components are known, it is simple to determine the hexagon sector by checking their sign. The procedure outlined in the flow chart of Fig. A9.2 accomplishes this task. The sequence of sign checks can be efficiently implemented with logic operations in the modulator routine. In the end, with a few lines of code we have determined (i) the position of the reference vector in the hexagon and (ii) the lengths of its projections on the two adjacent output voltage vectors (represented by one of the three computed $Z_{ix}Z_{iy}$ couples). We are therefore ready to program the PWM modulator to generate such vectors, plus one of the two possible zero vectors. There are only two final issues that need to be taken care of: the sequence of vector generation and the possible occurrence of saturation.

As far as the former issue is concerned, we present two examples of possible generation sequences in Fig. A9.3. Depending on the controlled system characteristics, one can be more advantageous than the other. As far as the latter is concerned, there is not a single straightforward way to cope with saturation.

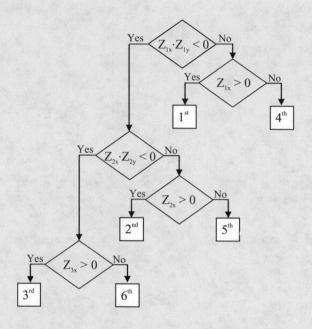

Figure A9.2: Flow chart of the sector identification algorithm.

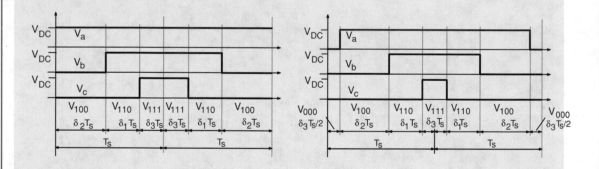

Figure A9.3: Two different application sequences for the same output voltage vectors. The sequence on the left implies a minimization of the number of switchings. The sequence on the right implies the minimization of the current ripple amplitude (voltage pulses have even symmetry). Note that each strategy develops in two adjacent modulation periods.

All strategies imply the acceptance of some degree of distortion of the output voltage. Once saturation is detected, which is easily done (the output vector durations, summed together, exceed the duration of the modulation period), some strategies reduce proportionally each duration until a sum equal to the modulation period duration is obtained. Other strategies consider the reduction of only one component (the shorter of the two) so as to get their sum again to be equal to the duration of the modulation period. The latter strategy, of course, implies the loss not only of the correct vector amplitude, but also of its phase. Another issue of some interest concerning saturation is the automatic change from linear to six-step modulation [5], which can be necessary in heavy saturation conditions. It is easy to verify that this is inherently achieved by the second saturation strategy we have just described.

Therefore, any time one of the inverter output voltage vectors is generated, a non zero γ component is produced on the load, that, being orthogonal to π is not visible in the vector decomposition of Fig. 5.4. Referring to Fig. 5.2, this means that SVM implies a particular modulation of the voltage between nodes N and G, V_{NG}. This is due to the common mode component of the inverter output voltage vectors. Indeed, it is easy to demonstrate that, in case of a symmetrical load structure, almost always encountered in practice, V_{NG} is instantaneously and exactly equal to the γ component of the inverter output voltage. The most important implication of this fact is that the phase to neutral voltage of the load will always be insensitive to any common mode component of the inverter output voltage, i.e., one can freely add common mode components to the $\vec{V}_{abc}$ vector, without perturbing the load voltage.

This is exactly what SVM implicitly does. Its effect, from the inverter's standpoint, can be proved to be very similar to that of third harmonic injection, sometimes employed in analog three phase PWM implementations. An increase by 15% of the voltage amplitude range that corresponds to a linear converter operation, i.e., to the absence of any saturation phenomenon, is obtained, as (5.12) clearly demonstrates.

This remark concludes our essential presentation of SVM. We are well aware that several other interesting issues could be addressed, but we feel like what we have presented is more than enough to allow us to discuss the following digital control application examples. The interested reader can find very useful additional information about SVM in the fundamental papers [3] and [4] and in several others that, in more recent times, have contributed to the development of PWM strategies for multi-phase converters.

5.2.1   SPACE VECTOR MODULATION-BASED CONTROLLERS

The typical organization of a three-phase VSI controller based on SVM is shown in Fig. 5.6. As can be seen, the controller takes advantage of the application of $\alpha\beta$ transformation to operate on two sampled variables instead of three. This not only simplifies the control algorithm, but also allows to directly generate the reference voltage components for the SVM in the $\alpha\beta$ reference frame. From those components, a suitable modulation procedure, like the one outlined in the Aside 9, will be able to determine the phase duty-cycles, managing inverter saturation if needed.

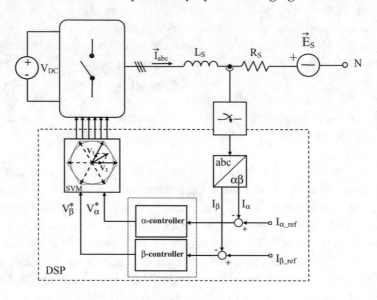

Figure 5.6: Organization of a three phase digital current controller based on SVM.

One could wonder whether the application of $\alpha\beta$ transformations to the controller input signals, in general, modifies the transfer function or state space model the controller design is based on. Clearly, if this is the case, passing from the three-phase system to the electrically equivalent two-phase one implies the need for a complete controller re-design. Luckily, this is hardly the case. Under the assumption of balanced and symmetrical load, we can indeed demonstrate that the design of the α- or β-axis controller is exactly identical to that of a single phase current controller operating on one of the three inverter phases. In order to show that, we need to define the continuous time state space model of the inverter and its load. It is easy to verify that this is given by:

$$
\frac{d}{dt}\begin{bmatrix} I_a \\ I_b \\ I_c \end{bmatrix} = -\frac{R_S}{L_S} \cdot \begin{bmatrix} 1 & 0 & 0 \\ 0 & 1 & 0 \\ 0 & 0 & 1 \end{bmatrix} \cdot \begin{bmatrix} I_a \\ I_b \\ I_c \end{bmatrix} + \frac{1}{3L_S} \cdot \begin{bmatrix} 2 & -1 & -1 \\ -1 & 2 & -1 \\ -1 & -1 & 2 \end{bmatrix} \cdot \begin{bmatrix} V_a \\ V_b \\ V_c \end{bmatrix}
$$

$$
- \frac{1}{L_S} \cdot \begin{bmatrix} 1 & 0 & 0 \\ 0 & 1 & 0 \\ 0 & 0 & 1 \end{bmatrix} \cdot \begin{bmatrix} E_{Sa} \\ E_{Sb} \\ E_{Sc} \end{bmatrix}, \tag{5.13}
$$

where the instantaneous neutral to ground voltage expression $V_{NG} = \frac{1}{3} \cdot (V_a + V_b + V_c)$ has been used. Now, if we apply to the different vectors in (5.13) the $T_{\alpha\beta}$ transformation, i.e., we replace each vector $\vec{x}_{abc}$ with $T_{\alpha\beta}^T \vec{x}_{\alpha\beta}$, after some re-arrangement, we get:

$$\frac{d}{dt}\vec{I}_{\alpha\beta} = -\frac{R_S}{L_S} \cdot T_{\alpha\beta} \cdot I_3 \cdot T_{\alpha\beta}^T \cdot \vec{I}_{\alpha\beta} + \frac{1}{3L_S} \cdot T_{\alpha\beta} \cdot \begin{bmatrix} 2 & -1 & -1 \\ -1 & 2 & -1 \\ -1 & -1 & 2 \end{bmatrix} \cdot T_{\alpha\beta}^T \cdot \vec{V}_{\alpha\beta}$$

$$-\frac{1}{L_S} \cdot T_{\alpha\beta} \cdot I_3 \cdot T_{\alpha\beta}^T \cdot \vec{E}_{S\alpha\beta}, \tag{5.14}$$

where I_3 is the 3×3 identity matrix. Simplifying the matrix products, we find the following result:

$$\frac{d}{dt}\vec{I}_{\alpha\beta} = -\frac{R_S}{L_S} \cdot I_2 \cdot \vec{I}_{\alpha\beta} + \frac{1}{L_S} \cdot I_2 \cdot \vec{V}_{\alpha\beta} - \frac{1}{L_S} \cdot I_2 \cdot \vec{E}_{S\alpha\beta}, \tag{5.15}$$

where I_2 represents the 2×2 identity matrix. Please note that the contribution of V_{NG} to the system dynamics, known as phase interference, has been canceled by the application of the $T_{\alpha\beta}$ transformation, as expected. Equation (5.15) shows that the equations for the two axes are now fully de-coupled, i.e., totally independent from each other. In addition, the structure and parameters of the two axes system are identical to that of the original three-phase system. Consequently, under the assumption of symmetrical and balanced load, it is not necessary to have any model adjustment and the design of the current regulator for the α and β axes can be done exactly as on a single-phase inverter.

This very important results implies that everything we have said about PI and predictive digital current control in the previous chapters, including the possibility of a multi-sampled implementation, can be immediately used also in three-phase inverters. The only additional elements we have to take into account are the implementation of a suitable SVM algorithm and of the $\alpha\beta$ transformation.

A warning must be issued in relation to the hysteresis controller. Its implementation in three phase systems, even a purely analog one, requires particular care, just because of the problem of phase interference. Indeed, while V_{NG} contribution is not essential in defining the *average* phase voltage and currents and can be eliminated with the the $T_{\alpha\beta}$ transformation, the same *cannot be done* for the *instantaneous* currents, that are the inputs for the hysteresis controller. In this case, the three wire inverter arrangement is such that switching one phase does not necessarily imply a slope reversal of the corresponding current, because, due to V_{NG} contribution, the instantaneous voltage across the phase inductor depends on the state of *all three* phases. This problem can actually be solved with some simple additional signal processing. Basically, an estimation of the V_{NG} contribution has to be subtracted from each measured current signal, making it independent from the others. The calculation can be easily performed by analog circuitry, as explained, e.g., in [6].

5.3 THE ROTATING REFERENCE FRAME CURRENT CONTROLLER

Once the three-phase inverter of Fig. 5.2 has been proved to be completely equivalent to a couple of independent single-phase inverters, other questions may be asked. Indeed, one could wonder whether the mapping of the system in the $\alpha\beta$ reference frame could be somehow exploited to *improve* the current controller dynamic performance.

   While this is not possible for the dead-beat controller, that already provides the best possible dynamic response among digital current regulators, in the case of the PI current controller the answer to the above question is affirmative. The implementation of the so called *rotating reference frame* controller indeed allows a significant improvement of the reference tracking capabilities of the PI regulator. This section is therefore dedicated to the illustration of the basic principles behind this solution.

   The first concept we have to introduce is that of Park's transformation, a very well-known tool for electrical machine designers.

5.3.1 PARK'S TRANSFORMATION

The idea behind Park's transformation is quite simple: instead of mapping the three-phase inverter and its load onto a fixed two axes reference frame, this new transformation maps it onto a two axes synchronous rotating reference frame. This practically means moving from a static coordinate transformation to a dynamic one, i.e., to a linear transformation whose matrix has time-varying coefficients.

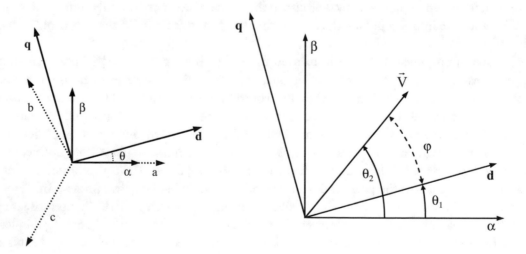

Figure 5.7: Vector diagrams for Park's transformation.

Before entering into the mathematical details, we may refer to Fig. 5.7 to get an idea of Park's transformation's meaning. The transformation defines a new set of reference axes, called d and q, that rotate around the static $\alpha\beta$ reference frame at a constant angular frequency ω. Referring to Fig. 5.7, this means that $\theta = \omega t$.

We have seen in Section 5.1 that the application of $\alpha\beta$ transformation to a triplet of symmetrical and balanced sinusoidal signals (5.7) turns them into a couple of 90° shifted sinusoidal signals (5.8), whose geometrical interpretation can be that of a rotating vector, $\vec{V}$. The rotating vector angular speed equals the angular frequency of the original voltage triplet, that we can consider the *fundamental* frequency of our three phase system. Now, if the angular speed of the rotating vector equals ω, what happens is that, in the *dq* reference frame, vector $\vec{V}$ is not moving at all! Referring again to Fig. 5.7, what we have just seen implies that angles θ_1 and θ_2 will both increase with angular frequency ω, while angle φ will be *constant* and so will be the lengths of vector $\vec{V}$ projections on the d and q axes.

The advantage of using the Park's transformation is represented exactly by the fact that sinusoidal signals with angular frequency ω, will be seen as *constant* signals in the *dq* reference frame. We have seen how a PI controller, especially a digital PI controller, can be affected by a non negligible tracking error with respect to sinusoidal reference signals, that is due to the limited closed loop gain at the frequency of interest. On the contrary, a PI controller can guarantee zero tracking error on constant signals, thanks to the built in integral action. Therefore, if a PI controller is implemented in the *dq* reference frame, without any additional provision, its tracking error with respect to sinusoidal signals having angular frequency equal to ω, i.e., to the frequency of Park's transformation, will become equal to zero. As we will see in the following, this principle is exploited in the implementation of the so called synchronous frame current controllers, where the Park's transformation angular speed is chosen exactly equal to the three-phase system fundamental frequency.

We can now show the mathematical formulation of Park's transformation. Considering Fig. 5.7, it is easy to demonstrate that this is given by the following matrix:

$$\begin{bmatrix} x_d \\ x_q \end{bmatrix} = T_{dq} \begin{bmatrix} x_\alpha \\ x_\beta \end{bmatrix} = \begin{bmatrix} \cos\theta & \sin\theta \\ -\sin\theta & \cos\theta \end{bmatrix} \begin{bmatrix} x_\alpha \\ x_\beta \end{bmatrix}, \tag{5.16}$$

where $\theta = \omega t$. Please note that, using the complex phasorial representation of vectors, (5.16) can be very simply expressed as:

$$\vec{x}_{dq} = x_d + j x_q = (x_\alpha + j x_\beta) \cdot (\cos\theta - j\sin\theta) = \vec{x}_{\alpha\beta} \cdot e^{-j\theta}. \tag{5.17}$$

It is easy to show that T_{dq} is associated to another orthonormal base of the $\mathbb{R}^2$ space, so that its inverse can be immediately found:

$$\begin{bmatrix} x_\alpha \\ x_\beta \end{bmatrix} = T_{dq}^T \begin{bmatrix} x_d \\ x_q \end{bmatrix} = \begin{bmatrix} \cos\theta & -\sin\theta \\ \sin\theta & \cos\theta \end{bmatrix} \begin{bmatrix} x_d \\ x_q \end{bmatrix}, \tag{5.18}$$

that, using again the complex phasorial notation, can be simply written as $\vec{x}_{\alpha\beta} = \vec{x}_{dq} \cdot e^{+j\theta}$.

As we did in the previous section, we can as well investigate the transformation of the system state equations, determined by the application of Park's transformation. To this purpose, all we need to do is consider (5.15) and use (5.18) on the left and right-hand sides of it. It is almost immediate to find the following result:

$$\frac{d}{dt}\vec{I}_{dq} = \begin{bmatrix} -\dfrac{R_S}{L_S} & +\omega \\ -\omega & -\dfrac{R_S}{L_S} \end{bmatrix} \cdot \vec{I}_{dq} + \frac{1}{L_S} \cdot I_2 \cdot \vec{V}_{dq} - \frac{1}{L_S} \cdot I_2 \cdot \vec{E}_{Sdq}, \qquad (5.19)$$

that shows a very interesting fact. The two-system dynamic equations are now complicated by the cross-coupling of the two axes, i.e., they are no longer independent from each other. This is the reason why, in control schemes like the one of Fig. 5.8, de-coupling feed-forward paths are sometimes included. These make the system dynamics totally identical to those of the original one.

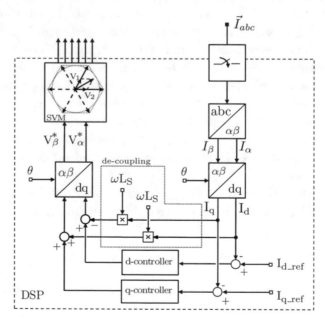

Figure 5.8: Organization of a three phase digital current controller in the *dq* reference frame.

To complete this brief discussion of Park's transformation we need to say that, in addition to what we have seen so far, it is also possible to implement the so-called *inverse sequence* Park's transformation. This is nothing but the transformation we have just presented, that we may now identify as the *direct sequence* Park's transformation, where the direction of the *dq* axes rotation is assumed to be *inverted*. It is immediate to verify that the implementation of the inverse sequence transformation simply amounts to swap the roles of (5.16) and (5.18).

One could wonder why the inverse sequence transformation is ever required, since we have shown that the direct sequence transformation is capable of mapping all the $\alpha\beta$ space without loss of information. The reason is that, so far, we have considered *balanced* and *symmetrical* three phase systems, but, more generally, impedance unbalances and/or unsymmetrical voltage sources can be found. In this case, a three phase system can be shown to be equivalent to the superposition of a direct sequence system and an inverse sequence system, both of them symmetrical and balanced and so both properly describable in the $\alpha\beta$ reference. If we neglect the so-called *omopolar* components, the superposition of *both* the direct and the inverse sequence two-phase systems is exactly equivalent to the original three phase system, while none of them is by itself. Of course, in case of zero or negligible unbalance/unsymmetry, the inverse sequence components will be accordingly zero or negligible, which motivates, in the majority of practical cases, the use of (5.16) and (5.18) alone.

Finally, it is important to underline that, because the elements of T_{dq} and T_{dq}^T are not time invariant, the application of Park's transformation, differently from the $\alpha\beta$ transformation, affects the system dynamics. This means that any controller, designed in the *dq* reference frame, is actually equivalent to a stationary frame controller that *does not maintain* the same frequency response. To keep the discussion reasonably simple, we refer, for the moment, to analog current regulators. In the end, we will see how to adapt our conclusions to digital current regulators.

5.3.2 DESIGN OF A ROTATING REFERENCE FRAME PI CURRENT CONTROLLER

For the reasons previously explained, we are very interested in PI controllers, that, once implemented in the rotating reference frame can offer zero steady state tracking error [7] for sinusoidal signals whose angular frequency is equal to ω. In some applications, where the phase error between the current reference signal and the inverter phase current must be as small as possible, this indeed represents the optimal solution. In order to properly design the PI controller in the *dq* reference frame, we need to understand what is the corresponding stationary frame controller. We begin by presenting a suitable model of the rotating reference frame current controller. This is shown in Fig. 5.9.

There are several important issues related to Fig. 5.9. First of all, as can be seen, the schematic is drawn in vector form, i.e., the indicated quantities represent bi-dimensional currents and voltages. Accordingly, the Park's transformation is simply represented by multiplication with the complex phasors $e^{\pm j\theta}$ where, as usual, $\theta = \omega t$. Secondly, both direct and inverse sequence transformation are taken into account, so as to make the schematic representative of as many practical cases as possible.

Even more importantly, the PI controller has been decomposed into a parallel structure, as it is always possible to do. Once this is done it is immediate to realize that, since the proportional gain K_P' is time invariant, the Park's transformation operators that would apply to it can be eliminated, as they are completely ineffective on a constant gain. Finally, the two proportional parts, operating respectively on the direct and inverse sequence components of the current error, can be unified, because they turn out to be exactly identical. This explains why a two factor multiplying the proportional gain, that, in principle, could have been omitted, has been explicitly indicated.

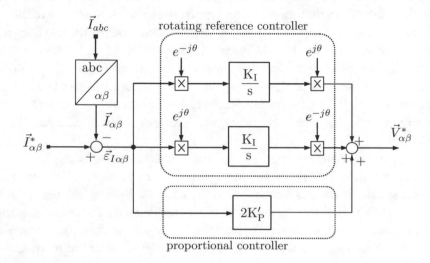

Figure 5.9: Organization of the rotating reference controller in the continuous time domain.

As can be seen, what we ended up with is a proportional controller, whose gain can be designed exactly as that of a single phase proportional current controller, that operates in parallel with two rotating reference frame integral controllers. The integral gain can be designed recalling that its effect will be to bring down to zero the tracking error with respect to sinusoidal reference signals having an angular frequency equal to ω. Of course, the higher the integral gain, the faster the achieved speed of response. An interesting problem, however, is how to pre-determine and control the settling time of the integral controller action, so as to avoid ringing, for example in the presence of a step reference variation. This problem can be effectively solved considering a different interpretation of the rotating reference controller, as is presented in the next section.

To conclude the discussion of rotating reference frame PI current controllers, we have to address the problem of its digital implementation. Of course, it is highly recommended that this solution is implemented digitally, as this makes it very simple to implement the different coordinate transformations involved in the controller operation. Once again, discretization is a very useful tool to accomplish this task. Based on what we have just seen, it is immediate to recognize that its application to the proportional part of the controller poses no significant problem. The only caution we need to apply may be in the continuous time domain design phase, where the phase margin we require for the open loop gain might be slightly oversized to cope with the calculation delay.

The application of discretization to the integral part of the controller is also relatively simple, because we have seen how Euler, or trapezoidal, numerical integration can effectively replace the analog integrator. The only caution we need to apply is the adjustment of the gain value, that has to multiplied by the sampling period. In conclusion, a possible schematic of the digital version of the controller presented in Fig. 5.9, is that shown on Fig. 5.10. Please note that the different vectors of Fig. 5.10 have now to be interpreted as sampled signals. It is also possible to see that the integral controllers have been discretized using the backward Euler method.

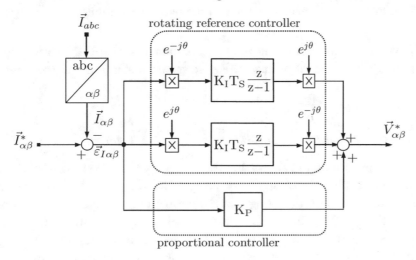

Figure 5.10: Discretized version of the rotating reference PI current controller, with $K_P = 2K'_P$.

5.3.3   A DIFFERENT IMPLEMENTATION OF THE ROTATING REFERENCE FRAME PI CURRENT CONTROLLER

We want now to derive an equivalent, stationary frame controller to replace the integral part of the rotating reference frame PI of Fig. 5.9. In order to do that, we consider the Laplace operator and, in particular, the following property:

$$\left[\mathcal{L} \left(e^{\lambda t} \cdot f\left(t\right) \right) \right](s) = \left[\mathcal{L}\left(f\right) \right](s - \lambda) \tag{5.20}$$

that is going to prove very useful to our purpose. Theorem (5.20) says that the multiplication by $e^{\lambda t}$ in the time domain results into a frequency translation in the s-domain. This means that, in the controller representation of Fig. 5.9, we can operate the substitution shown in Fig. 5.11.

Doing that, we obtain an equivalent stationary frame controller both for the direct sequence and for the inverse sequence components of the voltage reference vector, $\vec{V}^*_{dq+}$ and $\vec{V}^*_{dq-}$,

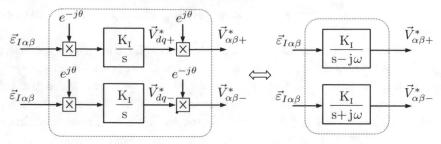

Figure 5.11: Laplace transformation of the rotating reference controller.

respectively. We then find, summing the two components, that the transfer function between the current error vector and the voltage reference vector, in the stationary reference frame, is the following:

$$\frac{\vec{V}^*_{\alpha\beta}(s)}{\vec{\mathcal{E}}_{I\alpha\beta}(s)} = \frac{\vec{V}^*_{\alpha\beta+}(s)}{\vec{\mathcal{E}}_{I\alpha\beta}(s)} + \frac{\vec{V}^*_{\alpha\beta-}(s)}{\vec{\mathcal{E}}_{I\alpha\beta}(s)} = \frac{K_I}{s-j\omega} + \frac{K_I}{s+j\omega} = 2K_I \frac{s}{s^2 + \omega^2}. \tag{5.21}$$

This very important result [8, 9] shows that the stationary frame equivalent of the rotating frame controller integral part is just a second order, resonant band pass filter, whose resonance frequency is exactly equal to ω. It is worth noting that the resonant filter presents zero damping factor and that the role of the integral gain is to determine the filter selectivity and, consequently, its settling time in response to perturbations. From (5.21) we see that increasing the K_I value determines a reduction of the filter selectivity and, consequently a faster settling time. On the contrary, reducing K_I determines a higher filter selectivity and, consequently, a longer settling time. A detailed explanation of the design criteria for this regulator, that allow to properly set the K_P and K_I gains, is reported in the Aside 10.

Aside 10. Design of a Stationary Frame Current Regulator with Zero Steady-State Error
    In Asides 2 and 3, we have determined the proportional and integral gains of a PI current controller. In this aside, we would like to illustrate a simple design example of a stationary frame current regulator composed, as shown in Fig. A10.1, of a proportional gain K_P and a single resonant controller $F_o(s)$:

$$F_o(s) = \frac{2K_I s}{s^2 + \omega_o^2}, \tag{A10.1}$$

tuned at the fundamental frequency ω_o. The considered parameter values are $V_{DC} = 250\,\text{V}$, $f_o = 60\,\text{Hz}$, $L_S = 3.5\,\text{mH}$, $R_S = 1\,\Omega$, $f_S = 10\,\text{kHz}$, $G_{TI} = 0.1\,\text{VA}^{-1}$. As done in Aside 2, the controller design is first performed in the analog domain and then translated in the

$\mathcal{Z}$-domain using a discretization process. The proportional gain K_P setting is based on the desired cross-over angular frequency ω_{CL}, as in any conventional PI control. Assuming that the current loop bandwidth is 1/10 of the switching frequency (i.e., $\omega_{CL} = 0.1\omega_S$), $K_P = 2K'_P \cong 2\omega_{CL}L_S/(2V_{DC}G_{TI}) = 0.88$. Instead, the integral gain K_I of the resonant regulator $F_o(s)$ is based on the desired transient response and on the specified phase margin ph_m. Indeed, since $\omega_{CL} \gg \omega_o$, $F_o(s) \approx 2K_I/s$, so that the design of K_I is the same as that of the PI controller of Aside 2, except for a factor 2. In our case, setting the phase margin $ph_m = 45°$, we have $K_I = K_{IN} = 1.25\,\text{krad}\,s^{-1}$. Figures A10.2 and A10.3 report the current reference, I_{OREF}, which is a sinusoidal waveform at ω_o, the output current I_O and the current error ε_I. As can be seen, the stationary frame current regulator is able to ensure zero steady-state errors for any reference or disturbance, whose frequency component is at ω_o.

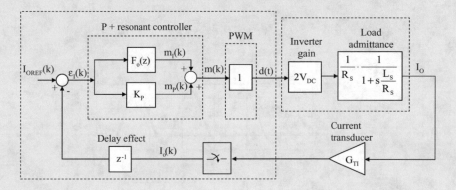

Figure A10.1: Block diagram representation of a stationary frame current regulator with zero steady-state error.

In order to highlight the properties of the resonant controller, we have reported in Fig. A10.4 the current control loop gain using three different integral gains: (a) $K_I = 2K_{IN}$, (b) $K_I = K_{IN}$, (c) $K_I = 0.1K_{IN}$. As can be seen, the integral gain K_I determines the filter selectivity and, consequently, its settling time in response to perturbations at the angular frequency ω_o; thus the higher the K_I, the lower the filter selectivity and, consequently, the faster the settling time. In contrast, the lower the K_I, the higher the filter selectivity and, consequently, the longer the settling time.

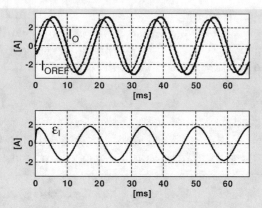

Figure A10.2: (top) Current reference I_{OREF} current I_O and (bottom) current error ε_I with a conventional PI control.

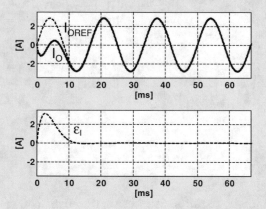

Figure A10.3: (top) Current reference I_{OREF} current I_O and (bottom) current error ε_I with a P+Resonant control.

In order to understand the settling time of the resonant controller and to establish an alternative second design criterion for the integral gain K_I, we may interpret the controller organization of Fig. A10.1 as a multi-loop scheme, where we first close the current control only with the proportional gain K_P. Then, the resonant filter $F_o(s)$ is designed so as to compensate the residual errors. From this point of view, the transfer function that the resonant controller $F_o(s)$ is going to compensate, once the proportional controller loop is closed, is

$$G_o(s) = \frac{m_I(s)}{e_I(s)} = \frac{1}{K_P} \underbrace{\frac{K_P G(s)}{1 + K_P G(s)}}_{W_P(s)} = \frac{1}{K_P} W_P(s), \qquad (A10.2)$$

where $W_P(s)$ is the transfer function between the current reference and the output current, when only the proportional controller is active. In general, $W_P(s)$ has a gain close to unity up to the desired bandwidth ω_{CL}. In our case, $W_P(s)$ is shown in Fig. A10.5. This controller interpretation leads to the following very interesting observations.

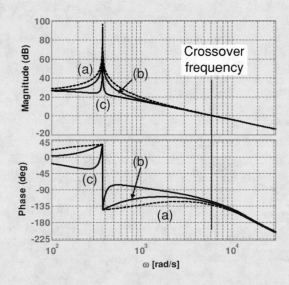

Figure A10.4: Current control loop gain for (a) $K_I = 2K_{IN}$; (b) $K_I = K_{IN}$; (c) $K_I = K_{IN}/10$.

(1) In the synchronous reference frame, the integrator controller K_I/s compensates a transfer function which is roughly approximated by $1/K_P$. Thus, the integral gain K_I can be designed given the desired cross-over frequency ω_{ro} (or desired time constant $t_{ro} = 1/\omega_{ro}$), i.e.

$$K_I = \frac{K_P}{t_{ro}} = 2.2 \frac{K_P}{t_r}, \qquad (A10.3)$$

where $t_r = n_o T_O$ is the desired response time (evaluated between 10% and 90% of a step response) for the fundamental frequency f_O. In our case, $t_r = 2\,\text{ms}$ or $n_o = 0.12$. Since we are reasoning in the synchronous reference frame, the time constant and the response time must be interpreted as related to the settling time experienced by the *envelope* of the fundamental frequency, f_O, component of the controlled current. The transient response in Fig. A10.3 is actually longer, because the step reference variation contains other frequencies, besides the fundamental one, f_O.

(2) Taking into account that any resonant controller determines a $+90°$ phase shift before the resonance frequency and $-90°$ phase shift after the resonance frequency, it is intuitive to understand that the resonant controller will be able to compensate only those frequencies f_O for which $W_P(j2\pi f_O)$ has a phase shift lower than $-90°$, so that the cross over of the $-180°$ stability limit is avoided. This imposes a limitation of the maximum angular frequency that it is possible to compensate, which must be kept lower than ω_L, as indicated in Fig. A10.5. This issue may be interesting for high-order harmonic compensation, as described in Aside 11.

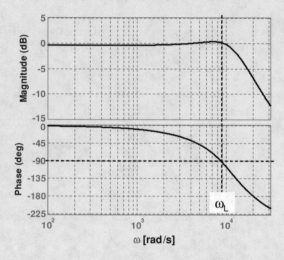

Figure A10.5: Bode diagram of $W_P(s)$.

The equivalence of the rotating reference frame PI controller with a proportional controller parallel connected to a tuned resonant filter suggests an alternative implementation of the controller that, not requiring the computation of Park's transformation, may offer a significant reduction of signal processing requirements for the control algorithm. Indeed, it is worth mentioning that the implementation of stationary frame resonant controllers, instead of synchronous reference frame controllers, has received, starting approximately from the year 2000, a significant attention from several research groups around the world, at least for those applications (UPS, PFC, active power filters, etc..) where the frequencies to be compensated are almost constant. Of course, the direct implementation in the discrete time domain of resonant filters with zero or very small damping factors, requires some care during the discretization process, in order to avoid warping effects that could shift the resonant frequencies, moving them out of the desired locations.

Aside 11. Stationary Frame Resonant Regulator: Extension to High-Order Harmonic Components and Introduction of a Phase Lead Compensation.

The approach presented in Section 5.3.3 can be extended to multiple harmonic compensation [9]. A typical example is the harmonic compensation in active power filters, where the current reference contains several harmonic components. The most straightforward approach for the compensation of the harmonic frequencies is the introduction of a resonant filter for each harmonic component to be compensated. Thus, referring to (A10.1), F_o becomes

$$F_o(s) = \sum_{k \in N_k} \frac{2K_{Ik}s}{s^2 + (k\omega_o)^2}. \qquad (A11.1)$$

N_k is the set of selected harmonic frequencies. Following the reasoning illustrated in the last part of Aside 10, K_{Ik} design is based on the transient response desired for each harmonic component. Thus, the design of each integrator gain K_{Ik} is given by

$$K_{Ik} = \frac{2.2K_P}{t_{r,k}} = \frac{2.2K_P}{n_{o,k}T_O}, \qquad (A11.2)$$

where $t_{r,k} = n_{o,k}T_O$ is the desired response time (evaluated between 10% and 90% of a step response) and $n_{o,k}$ is the number of fundamental periods T_O required by the settling of the generic k^{th} harmonic component of the controlled current. There is, however, a bandwidth limitation that applies to each harmonic component, given by angular frequency ω_L. Indeed, even for angular frequencies below ω_L, the transient response of the harmonic component may be lightly damped. As an example, using the parameters of Aside 10, we have set the harmonic component at 75% of ω_L (i.e., $k = 17$). The result is reported in Fig. A11.1, which clearly shows a lightly damped behavior.

This problem can be easily attenuated compensating the delay of the feedback loop by introducing a phase lead effect in the controller. As shown in Fig. A11.2, the phase lead ϕ_k is added when the outputs of the synchronous frame regulators $R_{kDC}(s)$ are transformed back to the stationary reference frame coordinates. Using Theorem (5.20), the relation between synchronous reference frame regulators $R_{kDC}(s)$ and stationary reference frame regulators $R_{kAC}(s)$ becomes

$$R_{kAC}(s) = \cos\phi_k \left[R_{kDC}(s - jk\omega_o) + R_{kDC}(s + jk\omega_o)\right]$$
$$+ j\sin\phi_k \left[R_{kDC}(s - jk\omega_o) - R_{kDC}(s + jk\omega_o)\right]. \qquad (A11.3)$$

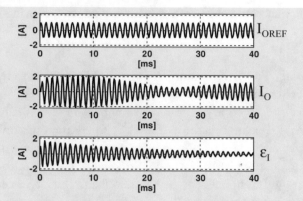

Figure A11.1: From top to bottom: current reference I_{OREF}, current I_O, and current error ε_I when the reference current is at $k\omega_0$ and a resonant filter tuned at harmonic k is used.

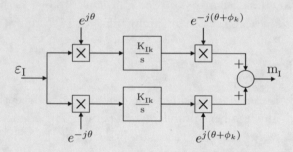

Figure A11.2: Rotating reference frame controller with phase lead ϕ_k.

which, for $\phi_k = 0$, corresponds to (A11.1). The leading angle ϕ_k can be set equal to the delay at frequency kf_O of the transfer function $W_P(s)$. The result of this provision is described in Fig. A11.3. Comparing this result with Fig. A11.1, we can clearly see the advantages of the introduction of a phase lead angle ϕ_k. If $R_{kDC}(s) = K_{Ik}/s$, (A11.3) yields

$$R_{AC}(s) = \sum_{k \in N_k} \frac{2K_{Ik}\left(s - \dfrac{\sin(\phi_k)}{\cos(\phi_k)}\right)\cos(\phi_k)}{s^2 + (k\omega_o)^2}, \tag{A11.4}$$

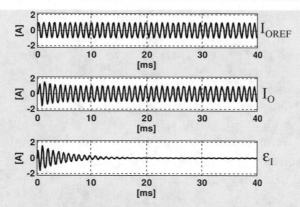

Figure A11.3: From top to bottom: current reference I_{OREF}, current I_O, and current error ε_I when the reference current is at $k\omega_0$ and a resonant filter tuned with lead angle ϕ_k at harmonic k is used.

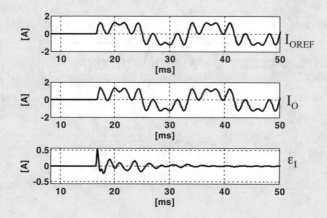

Figure A11.4: Transient response of a resonant controller $F_O(s)$: (top) current reference I_{OREF}; current I_O; (bottom) current error ε_I.

To further exemplify the performance of the proposed controller in a typical active filter application (please refer to Section 6.5.2 for the illustration of VSI applications as active power filters), we have simulated a reference signal I_{OREF}, which includes the fundamental component, the fifth and the seventh components, both with an amplitude equal to 50% of the fundamental one. Accordingly, we have implemented a resonant controller that includes the compensation of the fundamental fifth and seventh harmonic components.

The gain of the resonant controller has been set so as to have a response time equal to one fundamental period for all three harmonic components. The results are reported in Fig. A11.4. The figure shows how the residual error is reduced to zero after about one fundamental period, which is consistent with the specified dynamic response. As a comparison, we simulated an ideal dead-beat current controller, which ensures reference tracking with a two sample delay, reporting the results in Fig. A11.5. Note that, at the end of the simulated time interval, the residual error is still quite high, even if the dead-beat can be considered a very fast current controller. Of course, if higher-order harmonics were to be compensated (13th, 15th, etc.), the advantages of the resonant controller would be even greater than what the results reported in this example show.

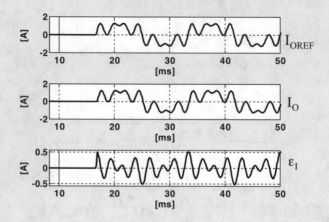

Figure A11.5: Transient response of an ideal dead-beat current controller: (top) current reference I_{OREF}; current I_O; (bottom) current error ε_I.

REFERENCES

[1] N. Mohan. T. Undeland, W. Robbins, "Power Electronics: Converters, Applications and Design," 2003, Wiley, ISBN 0–471-22693-9. 120

[2] J. Kassakian, G. Verghese, M. Schlecht, "Principles of Power Electronics," 1991, Addison Wesley, ISBN 02010–9689-7. 121

[3] H.W. Van Der Broeck, H.C. Skudenly, G.V. Stanke, "Analysis and Realization of a Pulsewidth Modulator Based on Voltage Space Vectors," *IEEE Transactions on Industry Applications*, Vol. 24, No. 1, January/February, 1988, pp. 142–150. DOI: 10.1109/28.87265. 121, 127

[4] D.G. Holmes, T.A. Lipo, "Pulse Width Modulation for Power Converters: Principles and Practice," IEEE Press Series on Power Engineering, 2003, ISBN 0–471-20814-0. 121, 127

[5] J. Holtz, W. Lotzkat, A. Khambadkone, "On Continuous Control of PWM Inverters in the Overmodulation Range Including the Six-Step Mode," International Conference on Industrial Electronics Control and Instrumentation (IECON), 1992, pp. 307–312. DOI: 10.1109/IECON.1992.254615. 127

[6] S. Buso, S. Fasolo, L. Malesani, P. Mattavelli: "A Dead-Beat Adaptive Hysteresis Current Control," *IEEE Transactions on Industry Applications*, Vol. 36, No. 4, July/August 2000, pp. 1174–1180. DOI: 10.1109/28.855976. 129

[7] D.N. Zmood, D.G. Holmes, "Stationary frame current regulation of PWM inverters with zero steady-state error," *IEEE Transactions on Power Electronics*, Vol. 18, No. 3, May 2003, pp 814–822. DOI: 10.1109/TPEL.2003.810852. 133

[8] D.N. Zmood, D.G. Holmes, G.H. Bode, "Frequency-domain analysis of three-phase linear current regulators," *IEEE Transactions on Industry Applications*, Vol. 37, No. 2, March/April 2001, pp. 601–610. DOI: 10.1109/28.913727. 136

[9] P. Mattavelli, "Synchronous-Frame Harmonic Control for High-Performance AC Power Supplies," *IEEE Transactions on Industry Applications*, Vol. 37, No. 3, May/June, 2001, pp. 864–872. DOI: 10.1109/28.924769. 136, 141

CHAPTER 6

External Control Loops

In previous chapters we have presented some examples of current control loop implementations, both for single and for three phase voltage source inverters. We have discussed how to design a PI current controller in the continuous time domain and how to turn it into a discrete time, or digital, controller. We also introduced the principles of dead-beat, predictive current control. In all these cases we have seen how the presence of a current control loop actually turns the VSI into a controlled current source with pre-determined speed of response and reference tracking accuracy. Exactly to maximize these parameters, we have also seen how to implement multi-sampled versions of the above mentioned controllers and measured the performance improvement they can offer.

However, there are several applications of VSIs where the implementation of a current control loop is just the first step to be taken. For example, in some cases, the control objective is not simply to develop a controlled current source, but rather to turn the VSI into a controlled *voltage* source. In other cases, the controlled current source is automatically regulated by an *external control loop* that is driven by another dynamic variable in the system, like, for instance, the rotational speed of an electrical motor. In these circumstances, the current loop becomes the inner control loop in a *multi-loop* arrangement of the VSI controller.

The purpose of this chapter is to present an overview of multi-loop control organizations, discussing some examples of external control loop design. Because a controlled current source can be used for a very large spectrum of different applications, it is practically impossible to deal with all. As we did before, also in this occasion we will limit our presentation to some typical application cases. In addition, we will limit the discussion to single phase VSIs, since we have shown, in the previous chapter, how the results can be almost directly applied to three-phase converters as well.

6.1   MODELING THE INTERNAL CURRENT LOOP

The set up of an external control loop around an existing current loop, typical of all multi-loop VSI control arrangements, poses questions similar to those we have been considering when, discussing the design of a current controller, we first tackled the static and dynamic modeling of PWM. Once again, independently from the nature and purposes of the external controller, its design requires a suitable model of the internal loop, taking into account static gain and dynamic response.

The derivation of this model is, in practice, simplified by the fact that, in deriving the current controller, all the involved transfer functions, associated to the different static and dynamic components of the system under consideration have been identified and calculated, even if, in some cases, under simplifying assumptions. From this standpoint, the designer's task is now easier, since he or she has to deal with a completely linearized dynamic system.

Any dynamic system analysis software allows to automatically calculate the closed-loop transfer function of a given feedback controlled system, once the various involved transfer functions are specified and their interconnection is suitably described by the user's program. This is indeed a very useful way of checking one's results, but we do not recommend this as a design approach. The problem is that the resulting dynamic model is typically high order, dependent on all the system parameters, and affected by all the approximations that were used in the derivation of the single transfer functions. Its practical usefulness for the design of the outer loop is therefore limited.

To effectively design the external loop what the designer actually needs is a first order simplified model of the internal loop, simple enough to be managed by pencil and paper calculations and, nevertheless, accurate enough to reproduce the system's dynamics with adequate detail. For this reason, in Chapter 4, we have stressed the importance of estimating the *phase lag* of the current control loop at a given frequency. When the frequency of interest corresponds exactly to the the cross-over frequency of the outer open loop gain, such phase lag represents the minimum required piece of information needed to complete the outer-loop design.

In the large majority of cases this simplified approach is sufficiently accurate to allow the successful design of any external loop. In some particular cases however, e.g., when the dynamic requirements for the external loop are demanding, the analytical, exact calculation of the internal loop response may be the only option available to the designer.

We can visualize the organization of a multi-loop digital controller considering Fig. 6.1. As can be seen, an additional dynamic variable, indicated as the external variable X_O, is introduced, that, after proper conditioning and sampling, is processed by a digital controller. The output of the external variable controller is the reference signal for the current controller, that is therefore driven by the external control loop. The shaded area in Fig. 6.1 represents the part of the system that is controlled by the current regulator and that, consequently, will be seen by the external loop as a single, lumped, dynamic system. Please note that this includes, as well, the holder delay effect embedded in the PWM modulator that, consequently, will not affect the external loop design.

The simplest modeling approach consists in the derivation of the block diagram of Fig. 6.2. As can be seen, the blocks appearing in Fig. 6.1 and pertaining to the external control loop are explicitly indicated; the current controller, the inverter and the load model are instead lumped in the current control block. Of course, to close the feedback loop, the definition of an additional transfer function, that relates the converter output current with the external control variable has to be specified as well.

While the latter transfer function depends on the particular application, and we will examine some particular cases in the following sections, the current control block model is independent of anything external to it. We can choose different model structures, based on the type of current

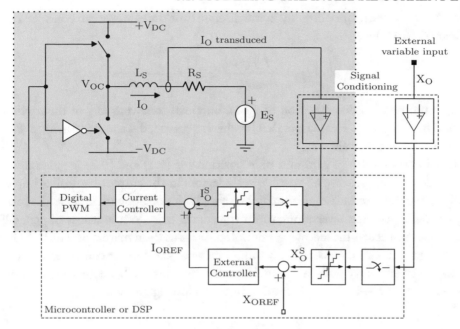

Figure 6.1: Typical organization of a multi-loop digital controller.

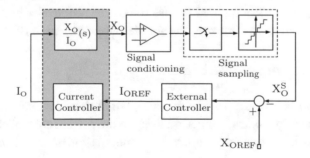

Figure 6.2: Block diagram of the external-loop digital controller.

controller we have actually implemented. An example, that is generally apt to model PI or other conventional regulators is the following:

$$W_I(s) = \frac{I_O(s)}{I_{OREF}(s)} \cong G_0 \frac{1}{1 + s\tau_{CC}}, \tag{6.1}$$

that can be used in case we want to proceed with the external controller synthesis in the continuous time domain and, later, apply some form of discretization. Please note that (6.1) is good also for a multi-sampled PI controller, with the appropriate choice of τ_{CC}. Instead, if the external controller synthesis has to be performed directly in the discrete time domain, we can consider a discrete time equivalent of (6.1), i.e.,

$$W_I(z) = \frac{I_O(z)}{I_{OREF}(z)} \cong \mathcal{Z}\left[G_0 \frac{1}{1 + s\tau_{CC}}\right], \tag{6.2}$$

that may represent the discretization of (6.1), obtained following any of the methods we have mentioned in Chapter 3. Please note that, in this latter case, the transfer function $X_O(s)/I_O(s)$ has to be discretized as well.

The determination of gain and pole position for (6.1) and (6.2) is generally simple. The gain depends on the presence of possible scale factors in the current controller implementation. Typically, when the internal variables are represented fractionally, with unity as the full-scale range value, gain G_0 equals the inverse of current transducer gain. Without loss of generality, we will assume this is the case in the following examples. As far as the dynamic part of (6.1) is concerned, the idea is again to simply model the response delay of the current control loop. If current loop design has been properly performed, (6.1) represents a reasonable approximation of the closed loop gain any time τ_{CC} is chosen according to the following relation:

$$\tau_{CC} = \frac{1}{2\pi f_{CL}}, \tag{6.3}$$

where f_{CL} represents the crossover frequency considered for the current loop.

A different approach can be used in case the current controller has been implemented as a digital predictive regulator. In that case, the simplest approximation of the current loop is represented by:

$$\frac{I_O(s)}{I_{OREF}(s)} = G_0 \frac{1 - sT_S}{1 + sT_S}, \tag{6.4}$$

where the static gain is identical to that in (6.2), while the dynamic term is a first-order Padè approximation of the two modulation period delay of the current controller. In the multi-sampled implementation, the delay would decrease to half a modulation period, determining the equivalent model:

$$\frac{I_O(s)}{I_{OREF}(s)} = G_0 \frac{1 - sT_S/4}{1 + sT_S/4}. \tag{6.5}$$

Of course, for the dead-beat current controller the discrete time model equivalent to (6.4) is:

$$\frac{I_O(z)}{I_{OREF}(z)} = \frac{G_0}{z^2}, \tag{6.6}$$

which, in this case, contains no approximations. Similarly, in the case of a multi-sampled implementation (see Section 4.2), the *exact* discrete time model is

$$\frac{I_O(z)}{I_{OREF}}(z) = \frac{G_0}{z}, \tag{6.7}$$

where the implicit sampling time is $T_S/2$, differently from (6.6) where it equals T_S. With the exception of (6.6) and (6.7), the modeling approaches we have just presented are fast and practical first order approximations of the current loop: therefore, it is always recommendable to verify their validity comparing them to a plot of the exact closed loop current control transfer function, calculated by any of the available dynamic system analysis software packages.

6.2   DESIGN OF VOLTAGE CONTROLLERS

A typical application field of VSIs is that of uninterruptible power supplies (UPSs). In this case, the voltage source inverter is used to implement a high quality, controlled voltage source. The technology of UPS systems involves a whole lot of other fundamental issues, like, for example, those related to energy storage and to the management of the interaction with the utility grid [1, 2]. Given the introductory purpose of this textbook, we will limit the discussion to some possible, and basic, strategies for the implementation of digital controllers of the UPS inverter stage. According to what we have illustrated in the previous section, the typical controller organization is multi-loop. The internal current control loop will be driven by an external voltage loop, as in the schematic diagram shown in Fig. 6.3 [3, 4].

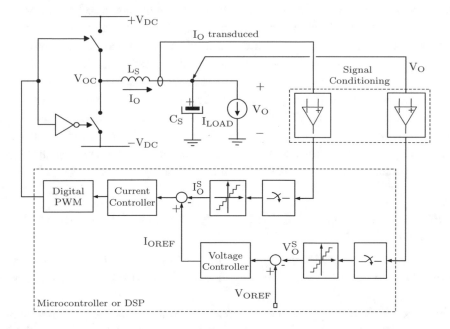

Figure 6.3: Typical organization of a single-phase UPS with digital control.

There are several aspects related to Fig. 6.3 that deserve further clarification. In the first place, the structure of the inverter output filter has been modified with respect to what we have been considering so far. The reason for this modification is that, in order to offer a relatively low impedance to external loads, schematically represented by a current source in Fig. 6.3, the converter output must be capacitive, at least in the frequency range of interest, that, in the case of UPS is set around the line frequency. In addition, the output capacitor provides, at least partially, load power factor correction, and gives to the UPS an energy storage capability to sustain the load, in the absence of the primary source of energy, for a predetermined amount of time, known as hold-up time. For the above reasons, the UPS inverter output filter will always have the configuration of Fig. 6.3.

It is worth noting that, in real cases, the load arrangement can be much more complex, e.g., including a transformer, so that the configuration of Fig. 6.3 represents just a simplified case study, that will allow us a relatively easier discussion of the basic control design aspects.

A second important issue related to the considered UPS system configuration is the motivation for the presence of a current control loop. One could observe that, provided the load structure is as shown in Fig. 6.3, there is actually no need for a current loop. The direct control of the output voltage could be implemented as is done, for instance, in DC-DC converters, when direct duty cycle control is sometimes used. This approach is, of course, perfectly possible and sometimes practically adopted too. Its main drawbacks are related to the protection of the inverter from accidental events like, for example, output short circuits. In this event, in order to avoid a potentially fatal overcurrent condition for the inverter switches, it is common practice to implement, at least, some current limitation mechanism, which, in turn, requires the implementation of current sensing and some form of signal processing, typically performed by analog circuitry. Therefore, there is not a significant cost reduction in the removal of the current loop. In addition, the presence of an internal current loop allows to decouple the second order output filter dynamics. This fact, differently from what one could expect, does not necessarily offer an advantage in the achievement of a faster dynamic response. However, the modularity, flexibility, higher tolerance to parameter variations and ease of design that characterizes multi-loop solutions make this the preferred strategy in commercial UPS designs.

6.2.1 POSSIBLE STRATEGIES: LARGE AND NARROW BANDWIDTH CONTROLLERS

The possible strategies for the implementation of a UPS output voltage controller can be roughly divided into two different categories: (i) large bandwidth controllers and (ii) narrow bandwidth controllers.

The large bandwidth approach is aimed at the *instantaneous compensation* of any deviation of the output voltage from its reference. A typical problem in UPS systems is the limitation of the output voltage waveform harmonic distortion within acceptable, product standard compliant levels. This is a particularly hard task when non linear, distorting loads, such as diode bridge rectifiers with capacitive output filter, are connected to the UPS output. Large bandwidth output voltage controllers try to achieve the goal by extending the regulation bandwidth so much to make it include a significant number of fundamental frequency harmonics (say 10, or even more). We will see in the following how this can be a very demanding control specification. Clearly, if this is achieved, the compensation of unwanted harmonic components of the output voltage will be achieved automatically, at least up to the regulator bandwidth. Typical implementations of this concept are linear PI regulators and dead-beat controllers. We will discuss both in the next section.

The narrow bandwidth approach is based on the following consideration. Examining the output voltage waveform distortion problem, one can realize that what is really needed is not the instantaneous compensation of all the undesired harmonic components. An harmonic compensation action that settles in a *few fundamental frequency periods* is actually enough to comply with product standards, provided that a relatively fast control of the fundamental harmonic component and a comparatively fast response to load variations is guaranteed. There are various possible implementations of this concept, ranging from repetitive based controllers to the adoption selected harmonic compensation by means of tuned filters. We will see some examples of this strategies in one of the following sections.

6.3   LARGE BANDWIDTH CONTROLLERS

This section is dedicated to the presentation of basic implementations of two output voltage control strategies for UPS systems, namely PI control and dead-beat control. The design approach, for both cases, closely resembles the one we have been following for the current controller implementation, where we have first met these types of regulator. Therefore, in the following, we will discuss in detail only the aspects that are peculiar to voltage controllers, being the generalities identical to those described in Chapter 3.

6.3.1   PI CONTROLLER

The implementation of a digital voltage PI controller is based on the general block diagram of Fig. 6.2, where we are now in the position of determining all the involved transfer functions. Prior to that, we need to summarize the main characteristics of the circuit of Fig. 6.3. We assume the UPS is built around our original case study VSI. The complete list of converter parameters is given by Table 6.1.

As can be seen, only some of the parameter values are the same originally reported in Table 2.1. Indeed, the output voltage specifications, relative to both amplitude and frequency, have been chosen, in the present case, so as to determine operating conditions that are typical of UPS systems in use in various non European countries around the world.

The design of the voltage controller requires the knowledge of the current controller dynamic characteristics. We can either assume that the current controller has been designed as in Chapter 2 and successively discretized or that we are dealing with a predictive current controller, of the type described in Chapter 3 and Chapter 4. One could point out that both these current controllers have been designed assuming a different inverter load configuration, in particular assuming the load voltage to be an exogenous input of the system and, as such, totally independent

Table 6.1: UPS inverter parameters

Rated output power, P_O	1500 (VA)
Phase inductance, L_S	1.2 (mH)
Output capacitor, C_S	85 (μF)
Output voltage, V_O	120 (V_{RMS})
Output frequency, f_O	60 (Hz)
DC link voltage, V_{DC}	250 (V)
Switching frequency, f_S	20 (kHz)
Current transducer gain, G_{TI}	0.1 (VA^{-1})
Voltage transducer gain, G_{TV}	0.02 (VV^{-1})

from the system's state variables (i.e., from the inverter output current I_O). It is immediate to see that, for the circuit of Fig. 6.3, this is no longer the case. However, it is possible to show that, for a typical UPS design, what we have seen in Chapters 2 and 3 is still valid and can be applied again.

A simple demonstration of this statement can be found in Fig. 6.4. The figure shows the Bode plot of the current control open loop gain, in the case of a PI controller designed exactly as outlined in the Aside 2. The plot is done both for the original load configuration (without capacitor) and for the new load configuration, including the output capacitor. The figure clearly demonstrates that, at the crossover frequency, and around it, the magnitude and phase of the two open loop configurations coincide. This is not at all accidental: in general, the output capacitor of a UPS inverter is sized to guarantee a certain (typically 50%) compensation of possible inductive loads (typical minimum load $\cos(\varphi)$ is 0.8), thus reducing the apparent load power the inverter has to generate. Because of that, differently from what happens in a DC-DC converter, in the UPS, the output capacitor is usually designed to operate at the *line frequency*. This implies that the inverter second order output filter has indeed a very low natural resonance frequency (about 500 Hz, in our example). This is extremely low with respect to the switching frequency, which implies that the filter impedance, close to the switching frequency, i.e., close to the typical desired crossover frequency of the current loop, is almost purely inductive. Therefore, designing a PI current controller on the second-order filter or designing it on the pure inductor, shorting the output capacitor, makes no practical difference.

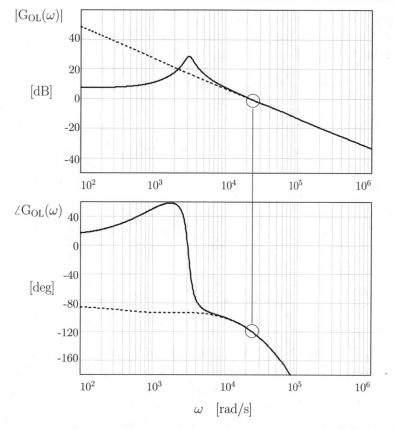

Figure 6.4: Bode plot of the current control open loop gain, with (solid line) and without (dashed line) output capacitor. The controller parameters are those calculated in Aside 2.

The case of the predictive controller requires a little more caution, but we will now show that the same conclusion can be reached. In order to do that, we consider a state space linear modeling of the second order filter (Fig. 6.3), that, recalling the Aside 5, can be simply represented in the following matrix form:

$$\frac{d}{dt}x(t) = A\,x(t) + B_1\,\overline{V}_{OC}(t) + B_2\,I_{LOAD}(t),\qquad(6.8)$$

where $x(t) = \begin{bmatrix} \overline{V}_O(t) & \bar{I}_O(t) \end{bmatrix}^T$ is the state vector, average inverter voltage $\overline{V}_{OC}$, and load current I_{LOAD} are considered system inputs, and

$$A = \begin{bmatrix} 0 & 1/C_s \\ -1/L_s & 0 \end{bmatrix} \quad B_1 = \begin{bmatrix} 0 \\ 1/L_s \end{bmatrix} \quad B_2 = \begin{bmatrix} -1/C_s \\ 0 \end{bmatrix}.\qquad(6.9)$$

Assuming, as we have done in Chapter 3, that the inverter voltage $\overline{V}_{OC}$ and load current I_{LOAD} are constant between sampling instants (zero-order hold equivalence of the system), the discrete time dynamic equations can be written as:

$$x(k+1) = \Phi\, x(k) + \Gamma_V\, \overline{V}_{OC}(k) + \Gamma_I\, I_{LOAD}(k), \tag{6.10}$$

where

$$\Phi = e^{AT_s} = \begin{bmatrix} \cos(\omega_o T_s) & \dfrac{1}{\omega_o C_S}\sin\omega_o T_s \\[2ex] -\dfrac{1}{\omega_o L_S}\sin\omega_o T_s & \cos\omega_o T_s \end{bmatrix} \approx \begin{bmatrix} 1 & \dfrac{T_s}{C_S} \\[2ex] -\dfrac{T_s}{L_S} & 1 \end{bmatrix}, \tag{6.11a}$$

$$\Gamma_V = \left(e^{AT_s} - I_2\right) A^{-1} B_1 = \begin{bmatrix} 1 - \cos(\omega_o T_s) \\[2ex] \dfrac{1}{\omega_o L_S}\sin(\omega_o T_s) \end{bmatrix} \approx \begin{bmatrix} 0 \\[2ex] \dfrac{T_s}{L_S} \end{bmatrix}, \tag{6.11b}$$

$$\Gamma_I = \left(e^{AT_s} - I_2\right) A^{-1} B_2 = \begin{bmatrix} -\dfrac{1}{\omega_o C_S}\sin(\omega_o T_s) \\[2ex] 1 - \cos(\omega_o T_s) \end{bmatrix} \approx \begin{bmatrix} -\dfrac{T_s}{C_S} \\[2ex] 0 \end{bmatrix}. \tag{6.11c}$$

In (6.11), I_2 is the 2×2 identity matrix, T_S is the sampling period and ω_o is the angular resonance frequency of the second order $L\text{-}C$ filter. Under the assumption that the sampling frequency is much higher than the resonance frequency of the $L\text{-}C$ filter (i.e., $\omega_o \cdot T_S \ll 1$), the approximations shown in (6.11a)–(6.11c), hold. Now, if we consider the second row of each matrix, we can immediately recognize that the current state equation implied by (6.11) is as follows:

$$\bar{I}_O(k+1) = \bar{I}_O(k) + \frac{T_S}{L_S} \cdot \left[\overline{V}_{OC}(k) - \overline{V}_O(k)\right], \tag{6.12}$$

which, once E_S is substituted by $\overline{V}_O$, is exactly coincident with (3.21). Once again, the predictive controller we can design around (6.12) is exactly the same we have designed around (3.21), which is then still valid both for its single-sampled and for its multi-sampled implementation.

In summary, thanks to the property of the considered topology that guarantees $\omega_o \cdot T_S \ll 1$, all we have said in Chapters 2, 3, and 4 is still valid and can be directly applied to the present case. Therefore, the design of the PI voltage controller can be developed assuming one of the solutions discussed in Chapter 3 or 4 is used in the current loop.

As an example, we are now going to discuss the case where the current controller is a conventional, i.e., single-update, dead-beat one. Of course, the same method we are now going to present can be applied in case a PI, or another kind of controller, is considered for the current loop.

We know, from Chapter 3, that the single-update dead-beat current controller is dynamically equivalent to a two modulation period delay. The static gain can be, without loss of generality assumed to be, from the voltage loop controller's standpoint, equal to the inverse of current transducer gain. Recalling the discussion of Section 6.1 and in particular (6.4) and Table 6.1, we can consider the transfer function for the closed loop current controller to be equal to:

$$\frac{I_O}{I_{OREF}}(s) = \frac{1}{G_{TI}} \frac{1 - sT_S}{1 + sT_S},$$
(6.13)

while that of the inverter load (Fig. 6.3) can be easily found to be given by:

$$\frac{V_O}{I_O}(s) = \frac{1}{sC_S}.$$
(6.14)

We can now build the block diagram around which the design of the PI voltage controller can be developed. This is shown in Fig. 6.5.

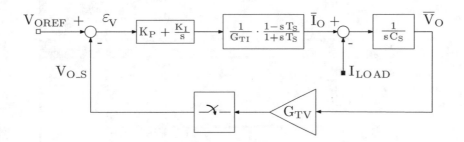

Figure 6.5: Block diagram of the voltage loop digital PI controller for the UPS of Fig. 6.3.

As can be seen, the control problem we are now considering is very similar to that considered in Chapter 2 for the continuous time PI current controller design. An important difference is in that the holder delay effect, for the reasons explained above, *has not to be considered* in this design.

The procedure to solve this problem, determining the PI controller gains K_P and K_I is presented in the Aside 12. As can be seen, it closely follows the one we considered in Chapters 2 and 3: first we determine a continuous time voltage PI controller that, later, we turn into a digital one by discretization. The PI voltage controller design is therefore concluded by the calculation of the discrete time equivalent of both gains. As usual, the final step we need to take is the simulation of the complete dynamic system, with current and voltage regulators. An example of the results obtained for our test case is shown in Fig. 6.6.

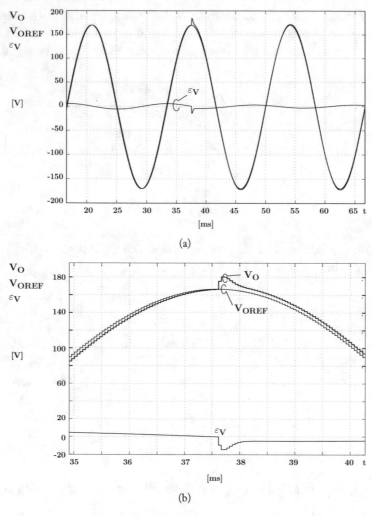

Figure 6.6: Dynamic response of the digital PI voltage controller: (a) response to a step-load discon-nection: the load current instantaneously reduces from about 9.5 A_{RMS} to 0 and (b) detail of previous figure.

When the PI controller is coupled to a faster current loop, such as the multi-sampled, double-update predictive controller presented in Section 4.2, the achievable performance can be significantly improved. This is due to reduction of the small-signal phase lag to a half of the modulation period, which, in turn, allows to push the bandwidth of the voltage loop much further. An example of the achievable performance, when an electronic load is connected at the inverter output and programmed to absorb a distorting current with crest factor $CF = 2$, is shown in

Fig. 6.7. As can be seen, the voltage waveform is practically not affected by the distorting load. Indeed, its measured THD, considering harmonic components up to order 40, is as low as 1.04%.

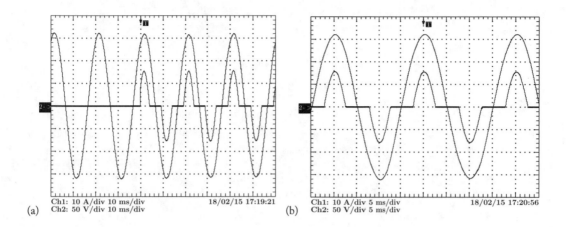

(a) Ch1: 10 A/div 10 ms/div   18/02/15 17:19:21
    Ch2: 50 V/div 10 ms/div

(b) Ch1: 10 A/div 5 ms/div   18/02/15 17:20:56
    Ch2: 50 V/div 5 ms/div

Figure 6.7: Dynamic response of the digital PI voltage controller with a faster inner current control loop: (a) response to a distorting load step connection and (b) steady-state operation with the distorting load. Please note that the output voltage frequency is $f_O = 50$ Hz in this experimental test.

Aside 12. Example of a PI Voltage Controller Design for a UPS Application

The voltage PI controller gains can be determined once the desired loop bandwidth, f_{CL}, is specified. For a UPS application, in order to achieve a satisfactory control of the voltage waveform in the presence of distorting loads, we can say that, as a rule of thumb, this should be at least 10–20 times the line frequency, i.e., from 600 to 1200 Hz in our example.

While this is easy to obtain when the switching frequency is relatively high, as it is in our case, and the current controller is a fast one, like the one we are considering here, in the opposite case, i.e., when a low switching frequency application is considered or when the internal control loop is relatively slow, it may not be too easy to achieve the desired f_{CL} values.

However, once f_{CL} is known, we can consider the open loop gain expression and force its magnitude to be equal to one at the desired crossover frequency. From Fig. 6.5, the open loop gain is found to be

$$G_{OLV}(s) = \frac{G_{TV}}{G_{TI}} \frac{1 - sT_S}{1 + sT_S} \frac{1}{sC_S} \left(K_P + \frac{K_I}{s} \right). \tag{A12.1}$$

It is worth noting that, differently from the current controller case, no delay effect related to the holder has been taken into account. This is possible because the internal current control loop has been designed to compensate for that. Therefore, the only dynamic delay the voltage controller has to compensate is that of the current controller.

Given (A12.1), the first condition we need to satisfy, by suitably choosing K_P and K_I, is as follows:

$$\frac{G_{TV}}{G_{TI}} \frac{\sqrt{K_I^2 + (\omega_{CL} K_P)^2}}{\omega_{CL}^2 C_S} = 1, \tag{A12.2}$$

where, as usual, $\omega_{CL} = 2\pi f_{CL}$. The second constraint we can impose is requiring a minimum phase margin, ph_m, for the loop gain at the crossover frequency. In order to get a reasonable damping of the dynamic response, this can be set equal to $60°$. Consequently, we find the following additional condition:

$$-180° + ph_m = -180° - 2\tan^{-1}(\omega_{CL} T_S) + \tan^{-1}\left(\omega_{CL} \frac{K_P}{K_I}\right). \tag{A12.3}$$

The solution of the system of Equations (A12.2) and (A12.3), considering the parameter values listed in Table 6.1 and imposing $f_{CL} = 720\,\text{Hz}$, provides us with the following values for the PI gains: $K_P = 1.92$, $K_I = 6.84 \times 10^2$ (rad s^{-1}).

The conversion of the continuous time PI into a discrete time one is simply obtained applying the following relations:

$$\begin{cases} K_{I\_dig} = K_I \cdot T_S \\ K_{P\_dig} = K_P. \end{cases} \tag{A12.4}$$

Finally, it is worth adding a comment on the calculation delay associated with the voltage controller. Typically, this can be considered equal to zero, because, if the controller hardware has been correctly chosen, the computation of the current reference sample can be done within the same control period where the duty-cycle is updated. In other words, it should always be possible to provide the current controller with the most recent sample of the current reference, without the need to wait for the following modulation period. The minimum requirement is, of course, that the sum of the durations of the voltage controller and current controller algorithms does not exceed one sampling period.

As can be seen, the steady-state reference tracking capabilities of the voltage controller are pretty fair. A steady-state sinusoidal tracking error is recognizable, that, as in the current loop case, is due to the finite amplification the PI controller offers at the reference frequency. This problem can be solved modifying the controller structure, as will be explained in Section 6.5, or attenuated by adding some form of feed-forward compensation, e.g., of the capacitive component of the inverter output current.

To test the voltage PI in dynamic conditions as well, we have considered a typical UPS test case, i.e., step-load disconnection. At the instant when the inverter output current is maximum, i.e., the maximum energy is stored in the L_S inductor, the load is disconnected. This causes an immediate output voltage error (negative) that needs to be corrected by the voltage controller. We can therefore evaluate the controller dynamic properties. It is worth noting that neither the current loop, nor the voltage loop, enters saturation during the test: accordingly, the behavior illustrated by Fig. 6.6 can be considered a consequence of the regulator properties, not influenced by saturation effects or other system non linearities. The regulation bandwidth determines the significant voltage error peak at the instant of the load step change. This is recovered in a relatively small fraction of the reference period, with adequate damping, i.e., without ringing or persistent oscillations.

6.3.2 THE PREDICTIVE CONTROLLER

In Chapters 3 and 4, we have been discussing the dead-beat, predictive current controller. We have seen how this represents a high-performance current controller, determining a dynamic response delay for the current loop that can be as low as half a modulation period. It may be quite obvious to ask if, using the same strategy, one could get the same high performance level for the voltage controller as well. The answer is affirmative: it is indeed possible to implement a predictive controller for the voltage control loop and get again a very fast dynamic performance. Following this approach, it is possible to set up a multi-loop controller based on decoupled current and voltage predictive regulators, whose dynamic response delay, when a conventional dead-beat controller is considered for the inner current loop, turns out to be equal to four modulation periods. This solution, that we identify as the multi-loop predictive controller, will be described in the next subsection.

However, for the sake of completeness, we have to say that the more direct and well-known application of dead-beat control to the converter structure of Fig. 6.3, does not actually follow the multi-loop approach. In this case, direct pole allocation and dynamic state feedback is applied to the second-order system described by (6.11). A multi-variable controller is consequently achieved, whose dynamic response delay is equal to three modulation periods, faster than the previously described one. However, as it almost always happens, the price to pay for the speedup is not negligible. The absence of a current control loop makes the management of some practical operating conditions, like overload or output short circuit, rather complicated. In the last part of this section, we will discuss the main features of this controller as well.

The Multi-loop Implementation

The schematic organization of the multi-loop predictive controller [5] is shown in Fig. 6.8. As can be seen, the block diagram is complicated by the presence of three functions, i.e., the capacitive current feed-forward (A), the reference current interpolator (B), and the load current estimator (C), that can be considered ancillary. As it will be explained a little further on, the purpose of these blocks is simply to improve the static and dynamic behavior of the regulator, but, for now, we can neglect them and focus on the main controller components.

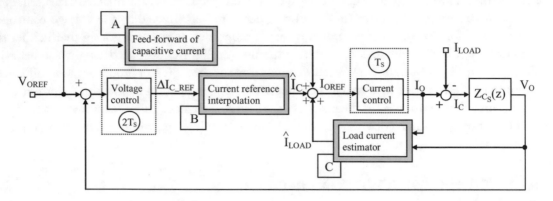

Figure 6.8: Schematic organization of the multi-loop predictive voltage controller.

Doing that, it is immediate to recognize in Fig. 6.8 the same basic organization of a multi-loop controller shown in Fig. 6.3. Of course, Fig. 6.8 is based on discrete time representations of both the controllers and the converter load. Because of that, no sampling block is explicitly represented in the figure. For the same reason, the load transfer function is represented as $Z_{C_S}(z)$, that stands for the discrete time version of (6.14).

Considering now the current controller, we will just say that, in this example, it is designed exactly following the procedure we described in Chapter 3. After the discussion of Section 6.3.1, we know that this is perfectly applicable to the present case, as the derivation of (6.12) clearly shows.

The voltage controller can be designed almost identically, considering the first row of the state space description (6.11). This corresponds to the following state equation

$$\overline{V}_O(h+1) = \overline{V}_O(h) + \frac{T_{S\_V}}{C_S} \cdot \left[\overline{I}_O(h) - I_{LOAD}(h)\right], \tag{6.15}$$

which, as can be seen, presents exactly the same structure of (6.12). Please note that, in order to keep the notation simple and clear, we denoted the considered sampling instant as $h \cdot T_{S\_V}$, to highlight that the sampling process for the voltage loop can be operated, in general, with a different sampling period, T_{S_V}, with respect to that of the current loop, T_S. Following the same reasoning presented in Chapter 3, we can now find the dead-beat control equation for the voltage loop. Once again, this presents exactly the same structure of the one derived for the current loop, i.e.,

$$I_{OREF}(h+1) = -I_{OREF}(h) + \frac{C_S}{T_{S\_V}} \cdot \left[V_{OREF}(h) - \overline{V}_O(h)\right] + 2 \cdot I_{LOAD}(h), \tag{6.16}$$

where the load current is assumed to be a relatively slowly varying signal and, consequently, the approximation $I_{LOAD}(h + 1) \cong I_{LOAD}(h)$ is considered.

It is essential to underline that the derivation of (6.16) actually hides a very important assumption, i.e., that it is possible and correct to *identify* the current reference signal with the actual inverter output current by the end of the *every given* control period. This assumption is not always correct: in particular, it is surely *not* correct if the sampling process for the voltage loop and that for the current loop have the same period duration. In this case, the dynamic delay of the current loop, that requires two periods to make the output current equal to its reference, undermines the system stability. On the contrary, if the sampling frequency for the voltage loop is set equal to one half of that used for the current loop, the delay, from the voltage loop standpoint, becomes ineffective and the identification of the current reference with the actual inverter output current is correct. Therefore, the controller organization of Fig. 6.8 actually requires $T_{S\_V} = 2T_S$. Because of that, the dynamic response delay of the voltage controller, that will be equal to two control cycles, as it was for the current controller, is actually equivalent to four modulation periods.

Several refinements are possible to improve the controller operation with respect to what can be achieved simply by programming (6.16) as the voltage loop control equation. In the first place, it is possible to feed-forward every known component of the inverter output current, like the current in the output capacitor C_S, that is easily pre-computed from the voltage reference signal, once the output capacitor value is known. This is exactly what block A of Fig. 6.8 does. The output current has another component, i.e., the load current, that, in general, cannot be pre-calculated, and, therefore, should be measured. Nevertheless, a simple estimation equation can be implemented, exactly as it was done for the current controller, in order to avoid this measurement, that can be sometimes problematic. The basic estimation equation is the following:

$$\hat{I}_{LOAD}(k - 1) = -\frac{C_S}{T_S} \cdot [\overline{V}_O(k) - \overline{V}_O(k - 1)] + \overline{I}_O(k - 1), \tag{6.17}$$

which can be actually improved by adding a cascaded low-pass filter, so as to remove possible instabilities or measurement noise. The implementation of (6.17) and of the low-pass filter is essentially the function of block C in Fig. 6.8.

Once the capacitive and load currents are obtained, thanks to blocks A and C, the function of the voltage controller is only to compensate the residual feed-forward and estimation errors. Of course, the voltage control Equation (6.16) can be rewritten accordingly, obtaining

$$\Delta I_C(h) = \frac{C_S}{2 \cdot T_s} \cdot [V_{OREF}(h) - \overline{V}_O(h)] - \Delta I_C(h - 1), \tag{6.18}$$

which explains why, in Fig. 6.8, the output of the voltage controller is not I_{O_REF}, but the quantity ΔI_{C_REF}.

The function of block B is a little more complicated to explain. We have seen before that the voltage control equation is computed at half the frequency of the current control. This means that the current controller reacts to the reference signal generated by the voltage controller as to a stepwise function, updated every two modulation periods. This determines persistent high frequency oscillations in the inverter output current. In order to eliminate this effect, the interpolator block B of Fig. 6.8, generates an extra reference signal sample to feed the current controller in those control periods when the voltage loop would not update the reference. This makes the reference signal for the current controller practically equivalent to a continuous time signal, correctly sampled with T_S period, and thus eliminates the step response dynamics from the output current.

The provisions we have briefly outlined make the UPS controller of Fig. 6.8 quite effective. We can see the typical performance achievable with this controller in Fig. 6.9. It is interesting to compare Fig. 6.9 and Fig. 6.6, since they were obtained for the very same test conditions.

As can be seen, there is a significant difference in the two controllers' performance. The dead-beat voltage controller presents a certain, residual steady-state tracking error at the fundamental frequency. The output voltage is, indeed, a delayed replica of the reference, which generates an instantaneous error at the fundamental frequency. Because of the relatively long sampling period for the voltage loop, the error is not negligible. On the other hand, the dead-beat controller's dynamic performance is much faster than the PI's, as is clearly visible if one compares the error trajectory after the load transient. This readiness guarantees both a smaller voltage overshoot and a faster recovery of the nominal voltage trajectory.

The Multi-variable Implementation

The dead-beat controller is more often implemented as in [6, 7, 8], i.e., by applying state feedback theory to the second-order dynamic system represented by (6.11). The approach practically replicates the one we have been following in the Aside 5, with the remarkable difference that we are now dealing with a two-component state vector. We can describe the solution considering, at first, the simpler and ideal case where the computation delay is neglected. Accordingly, the basic static state feedback implementation is the following:

$$x(k + 1) = \Phi \cdot x(k) + \Gamma_V K \cdot x(k) = \Phi_F \cdot x(k), \qquad (6.19)$$

where the system input $\overline{V}_{OC}$ has been replaced by $K \cdot x$, and $K = \left[K_{\overline{V}_O} K_{\overline{I}_O} \right]$ is the feedback gain matrix. Consequently, the closed-loop system is now characterized by a new state matrix $\Phi_F = \Phi + \Gamma_V K$, whose eigenvalues can be properly allocated by suitably choosing the gains $K_{\overline{V}_O}$ and $K_{\overline{I}_O}$. The computation is a little involved, but it is possible to see that the following values:

$$k_{\overline{V}_O} = \frac{1 - 2 \cos(\omega_o T_s)}{2 - 2 \cos(\omega_o T_s)} \quad k_{\overline{I}_O} = -\omega_o L_S \frac{1 + 2 \cos(\omega_o T_s)}{2 \sin(\omega_o T_s)}, \qquad (6.20)$$

achieve the desired result, i.e., both the closed-loop system eigenvalues are re-located in the origin of the complex plane. It is interesting to note that (6.20) is given for the original system matrices, i.e., without any approximation. It has therefore general validity. More subtly, if we tried to operate the closed-loop compensation of the system from the $\overline{V}_{OC}$ input, *after* the system is dynamically

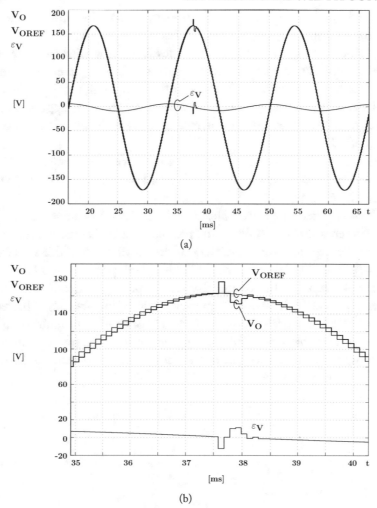

Figure 6.9: Dynamic response of the digital dead-beat voltage controller: (a) response to load step disconnection, (b) detail of previous figure.

de-coupled and the approximated system matrices are obtained, we would encounter a serious problem: the approximated dynamic system is no longer state *controllable* from the $\overline{V}_{OC}$ input. This reflects the physical fact that, in the hypothesis of a decoupled system, the output voltage $\overline{V}_O$ no longer depends on the average inverter voltage, but only on the average inverter current.

Therefore, the approach we are here discussing is only meaningful if we don't take into account the dynamic decoupling hypothesis. Please note that this could be the only correct way of synthesizing a dead-beat controller in all those cases where the second-order output filter does not guarantee the condition $\omega_o \cdot T_S \ll 1$ is satisfied.

In conclusion, the organization of a state feedback loop with gains given by (6.20) guarantees a dead-beat response for the closed-loop system. Unfortunately, the practical implementation of this solution is not possible, because of the computation delay, that we have not taken into account. In order to do that, we need to follow again the same approach of the Aside 5, i.e., considering a dynamic state feedback implementation. The details of the procedure are given in the following Aside 13.

Before we conclude our presentation of the multi-variable dead-beat controller, we would like to discuss the results of its numerical simulation, shown in Fig. 6.10. In particular, we would like to compare Fig. 6.10 and Fig. 6.9.

As can be seen, there is a certain, but not dramatic, performance improvement in the last considered implementation, especially in the residual steady-state tracking error, while the transient response is not significantly better than in the previous case. This is because, although, in principle, the multi-variable dead-beat controller is capable of a three modulation period response delay, i.e., the fastest theoretically possible dynamic response, the need for the reference signal reconstruction, as explained in the Aside 13, partially cancels this advantage. Therefore, the achieved dynamic performance is practically comparable to that offered by the multi-loop dead-beat implementation. However, we need to highlight, once again, that the multi-variable implementation is actually the only possible solution for a dead-beat control of second order output filters with relatively high resonance frequency.

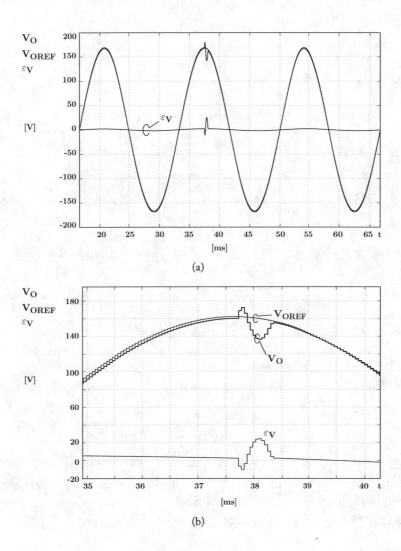

Figure 6.10: Dynamic response of the digital multi-variable dead-beat voltage controller: (a) response to a load step disconnection and (b) detail of previous figure.

Aside 13. State Feedback Derivation of the Predictive Voltage Controller

We consider the discrete time equivalent model for the UPS system given by (6.11), which we recall here for clearness, i.e.,

$$x(k+1) = \Phi \cdot x(k) + \Gamma_V \cdot \overline{V}_{OC}(k) + \Gamma_I \cdot I_{LOAD}(k), \tag{A13.1}$$

where $x(k) = [\overline{V}_O(k) \quad \overline{I}_O(k)]^T$ and

$$\Phi = \begin{bmatrix} \cos(\omega_0 T_s) & \dfrac{1}{\omega_0 C_S}\sin(\omega_0 T_s) \\ -\dfrac{1}{\omega_0 L_S}\sin(\omega_0 T_s) & \cos(\omega_0 T_s) \end{bmatrix}, \quad \Gamma_V = \begin{bmatrix} 1 - \cos(\omega_0 T_s) \\ \dfrac{1}{\omega_0 L_S}\sin(\omega_0 T_s) \end{bmatrix},$$

$$\Gamma_I = \begin{bmatrix} -\dfrac{1}{\omega_0 C_S}\sin(\omega_0 T_s) \\ 1 - \cos(\omega_0 T_s) \end{bmatrix}. \tag{A13.2}$$

What we want is to build a dynamic state feedback controller around variable $\overline{V}_{OC}$, which can be represented by the following equation:

$$\overline{V}_{OC}(k+1) = [K_1 \quad K_2] \cdot (x_{REF}(k) - x(k)) + K_3 \cdot \overline{V}_{OC}(k), \tag{A13.3}$$

where gains K_1, K_2, and K_3 have to be determined and $x_{REF}(k) = [\overline{V}_{OREF}(k) \quad \overline{I}_{OREF}(k)]^T$ is the state reference trajectory. We can now determine the augmented state matrix that corresponds to the new dynamic system, made up by (A13.1) and (A13.3). It is immediate to find that this is given by

$$\Phi_A = \begin{bmatrix} \Phi_{11} & \Phi_{12} & \Gamma_{V11} \\ \Phi_{21} & \Phi_{22} & \Gamma_{V21} \\ -K_1 & -K_2 & K_3 \end{bmatrix}. \tag{A13.4}$$

As we did in Aside 5, we now need to calculate the K_1, K_2, and K_3 gain values, so as to force the eigenvalues of matrix Φ_A to move to the origin of the complex plane. Once again, this very simple idea requires some mathematics; however, after that, it is possible to find that the following values

$$K_1 = -\frac{1 + 2\cos(\omega_0 T_S) - 4\cos^2(\omega_0 T_S)}{2[1 - \cos(\omega_0 T_S)]},$$

$$K_2 = -\frac{\omega_0 L_S}{2\sin(\omega_0 T_S)}[1 - 2\cos(\omega_0 T_S) - 4\cos^2(\omega_0 T_S)], \qquad \text{(A13.5)}$$

$$K_3 = -2\cos(\omega_0 T_S),$$

solve the problem. Therefore, substituting (A13.5) gains into (A13.3) control equation, we get the desired multi-variable dead-beat controller.

It is important to underline that, differently from the multi-loop implementation, in the multi-variable dead-beat controller, the computation of the current reference trajectory is not automatic, i.e., inherent in the controller structure. This means that we have to explicitly determine the reference current from the voltage reference trajectory, that is, of course, given, and from other system variables. The standard procedure is to pre-compute the capacitive current component of the output current from the voltage reference and either measure or estimate the load current. Estimation techniques, e.g., based on *disturbance observers* [9, 10], can be implemented that allow one to save the load current measurement. However, in that case, the observer dynamics are responsible for a certain increase in the response delay of the controller.

We conclude this brief discussion of dead-beat voltage controllers observing that, in recent times, a significant research effort has been focused on this control technique. Therefore, several technical papers can be found where this subject is treated in detail and possible refinements or different implementation strategies are presented. The interested reader may take advantage of references [6, 7, 8, 9, 10] as far as the multi-variable implementation is concerned. Instead, additional details on the multi-loop dead-beat controller implementation can be found in [5].

6.4   NARROW BANDWIDTH CONTROLLERS

In this section we present a summary review of two very popular *narrow bandwidth* voltage control strategies, frequently employed in UPS systems. These are the repetitive-based voltage controller and, once again, the rotating reference frame voltage controller. The former is based a totally new concept we never encountered before, the latter, instead, is the almost direct extension of what we have been discussing, in Chapter 5, for the current loop implementation. Essentially for this reason, we will here discuss a different implementation strategy for the same concept, that is based on DFT filters.

The Repetitive-based Voltage Controller

The concept of repetitive control originates from the internal model control principle. For obvious reasons, we are not going to present here any of the numerous theoretical issues related to internal model control and, in particular, to the derivation, under general assumptions, of repetitive controllers. The interested reader can find a very good treatment of these topics in specialized textbooks, like, for example, [11]. Instead, we would like to open our discussion simply by describing the goal of any repetitive controller, that is to make the controlled system output track a set of pre-defined reference inputs, without steady-state error. The theory shows that, in general, the achievement of this result requires the stabilization of an augmented system, where the dynamic representation, in terms of Laplace or $\mathcal{Z}$-transform, of the reference signal of interest, has been somehow added to the original system model. This can be, in some cases, a quite complicated control problem.

However, in the particular case of sinusoidal reference signals, that represent exactly what we are interested in, for the UPS output voltage control, the digital implementation of a repetitive controller becomes relatively simple, requiring only the set up of a suitably sized delay line and of a positive feedback loop [12, 13].

An example of the basic structure of a repetitive controller, organized for application to the UPS external voltage loop, is shown in Fig. 6.11(a). According to the required control function, the error on the UPS output voltage, ε_V, represents the controller input, while the controller output is represented by the current reference signal for the internal current loop.

It may not be obvious to see why, once the closed-loop system is stabilized, the configuration of Fig. 6.11(a) necessarily implies zero reference tracking error with respect to sinusoidal signals. The formal way to realize why and how this happens consists in computing the transfer function that relates the controller input to the output and plot the frequency response. What can then be found is a very interesting result: the controller transfer function presents infinite gain at all frequencies that are integer multiples of a fundamental one. The fundamental controller frequency is the one associated to the delay line duration. Therefore, if the delay line duration is made equal to the desired output voltage frequency, the frequency response of the repetitive controller will be approximately equivalent to the parallel connection of a bank of resonant filters, each presenting infinite gain at one integer multiple of the output voltage frequency.

As a matter of fact, this result can also be anticipated simply referring to Fig. 6.11(a) and considering the delay line operation. Any signal that repeats itself exactly in the delay line period, gets infinite amplification. Therefore, all sinusoidal signals, whose period is an integer sub-multiple of the delay line period, $M \cdot T_S$, get infinite amplification. One way or the other, we see that the controller structure of Fig. 6.11(a) is a practical means to boost to infinity the open-loop system gain at every harmonic up to the Nyquist frequency. From this necessarily derives a zero steady-state tracking error on the output voltage sinusoidal signal and on all of its harmonics.

However, exactly for the same reason, this structure poses serious stability problems for the system. Indeed, the infinite amplification of the highest order harmonic components of the

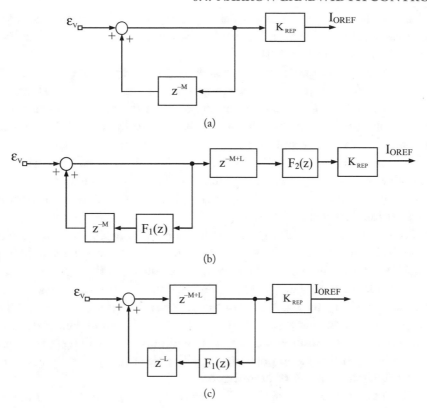

(a)

(b)

(c)

Figure 6.11: (a) General implementation of the repetitive controller, (b) provisions to improve the stability margin, and (c) the considered implementation.

voltage error can reduce the control loop phase margin and undermine the controller stability. The basic reason is that, as we know, the internal current controller has a limited bandwidth. Therefore, in order not to incur in instability, the frequency content of the current reference signal has to be limited accordingly.

Because of that, several additional provisions have been proposed for an effective practical implementation of the repetitive controller. For example, in order to guarantee system stability, some filters can be introduced in the scheme of Fig. 6.11(a), in the feedback path, $F_1(z)$, or in a cascade connection with the repetitive controller, $F_2(z)$, or even both, as shown in Fig. 6.11(b). The goal of these filters is exactly to limit the amplification of the high-order harmonics. In addition, the stability of the repetitive controller has been shown to greatly improve if a delay line of $M - L$ samples is inserted at the output of the regulator. This is actually equivalent to adding a phase lead of L samples for all the harmonic frequencies and has been shown [12] not to change the gain at the harmonic frequencies, but just to increase the system phase margin.

In conclusion, the repetitive controller organization we are going to discuss, that sums up all these considerations, is shown in Fig. 6.11(c), which is, of course, theoretically equivalent to the scheme of Fig. 6.11(b) when $F_2(z) = 1$.

In recent times, a lot of different voltage loop controllers, built around the repetitive controller structure of Fig. 6.11(c), have been proposed and applied. The different solutions try to solve the typical problems that are often encountered in the practical application of repetitive controllers. In particular, experience shows that it is normally quite difficult to achieve simultaneously a satisfactory steady-state voltage error compensation and an acceptable large-signal behavior from the repetitive controller in stand alone configuration. Stability can be obtained, but, due to the effects on the control loop phase of the high-frequency resonances in the controller frequency response, the phase margin is typically low, with a consequent unsatisfactory performance during transients.

For this reason, the repetitive controller is more typically employed in parallel connection with a conventional regulator. In the scheme of Fig. 6.12, we can see a simple implementation of this principle: a purely proportional controller is paralleled to the repetitive one. The motivation for the considered controller's organization is to have, in the steady state, the proportional controller action joined by the repetitive controller's one: the latter compensates the periodic error components the former, because of its limited bandwidth, cannot eliminate, thus making the residual tracking error practically equal to zero. In addition, as we will see, the solution allows the designer to better control the loop phase margin. Therefore, it is generally possible to guarantee a conveniently damped response to perturbations.

Seen from this standpoint, the repetitive controller can be considered as an optional function we can employ in parallel to a conventional controller any time we need to improve its steady-state performance. In the presence of periodic output voltage disturbances, like those induced by non linear loads connected to the UPS output, this solution can greatly improve the quality of the output voltage regulation. Of course, nothing can be gained from this controller organization in the compensation of fast transients, like those determined by step load variations.

The design of the parallel structure of Fig. 6.12 can be performed in two separate steps: (i) design of the proportional regulator and (ii) design of the repetitive controller. The first step is very similar to the standard PI design we have already described in Section 6.3.1 and Aside 12, so we will not comment further on that. As far as the second step is concerned, we basically need to determine: (i) the value of parameter M, (ii) the value of parameter L, (iii) the value of gain K_{REP}, and (iv) the structure of $F_1(z)$.

The design of parameter M simply requires the determination of the ratio between the sampling frequency and the fundamental output voltage frequency. Since M must be integer, this may generally require the adjustment of the switching frequency to an integer multiple of the output voltage fundamental. In our test case, the switching and sampling frequencies were adjusted to 19.92 kHz, thus giving $M = 332$.

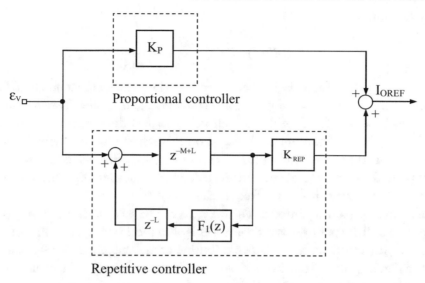

Figure 6.12: (a) Suggested repetitive-based voltage controller. The repetitive controller structure of Fig. 6.11(c) is parallel connected to a conventional, purely proportional controller.

The design of the other parameters requires the careful consideration of the open-loop gain, and in particular of the system phase margin. In order to compute the loop gain, we can refer to the block diagram of Fig. 6.13, where, once again, the basic organization of Fig. 6.5 can be identified, with the important difference that all blocks are now discrete time and, consequently, the ideal sampler block is no longer represented.

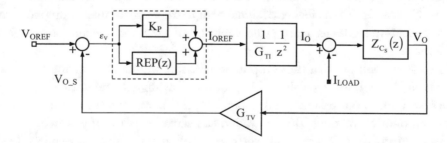

Figure 6.13: Repetitive-based voltage control loop. The scheme is used for the computation of the open-loop system gain.

As described above, the repetitive-based controller is given by the parallel connection of the purely proportional regulator and the repetitive controller of Fig. 6.11(c), whose transfer function can be easily found to be equal to

$$REP(z) = K_{REP} \frac{z^{-M+L}}{1 - z^{-M} F_1(z)}. \tag{6.21}$$

In addition to this, Fig. 6.13 includes the current loop transfer function that, being supposed to be of dead-beat type, is given by the usual static gain and an ideal two-period delay transfer function. Finally, as we already did, we indicate by $Z_{C_S}(z)$ the discrete time version of (6.14), obtained by any discretization method. Based on this scheme, we can now compute the open-loop gain and suitably select the repetitive controller parameters so as to maintain the system phase margin and crossover frequency unaffected, while achieving a significant gain boost at least for the first output voltage harmonic frequencies.

The open loop gain is plotted in Fig. 6.14. As can be seen, with the chosen parameters, the open loop gain of the repetitive-based controller is asymptotically equal to that of the purely proportional one. The repetitive controller contribution on the magnitude is represented by the gain peaks, located at integer multiples of the output voltage fundamental frequency and by the small increase of the equivalent proportional gain that appears as an offset between the two plots. The amplitude of the peaks has been limited in high frequency by using, as $F_1(z)$, a moving average filter with 21 taps. This, together with a suitable choice of parameter K_{REP}, that, in our example, has been set equal to 0.8, has allowed to achieve a phase margin at the crossover frequency that is practically identical to that of the purely proportional controller, thus avoiding any stability problem. In addition, no phase lead action was needed in the example we are here considering, since the sampling frequency is relatively high with respect to the crossover frequency. Finally, the effect on the loop phase determined by the moving average filter has been compensated by reducing by 10 the number of taps in the delay line. This provision is required because the 21 tap moving average filter actually gives a contribution to the loop phase that is equal to that of a 10 tap delay line. Therefore, the length of the delay line has to be reduced accordingly, so as to keep the total phase lag of the feedback signal path to the correct value. If this were not done, the frequency allocation of the resonant peaks could be affected and so could be the effectiveness of the regulator.

One could point out that the computational effort required for the implementation of this regulator is relatively high, typically calling for a not negligible amount of hardware resources. We have seen that, in our example, a 332 tap delay line is *theoretically* required, which implies a significant amount of memory. This limitation can actually be partially overcome by using a M_c sample decimation factor, thus reducing the number of taps the delay line requires. In the example reported hereafter, $M_c = 4$ and, consequently, the number of delay line taps M has been reduced to 83, i.e., to 81 to take the moving average filter into account. Indeed, the moving average filter $F_1(z)$ has been reduced to only 5 taps. Using this decimation factor the dynamic performance was not significantly affected. One issue related to the adoption of sample decimation is that the output of the repetitive control is updated only every M_c samples and is seen by the proportional controller as a stepwise function. Thus, an interpolator (first-order-hold, low-pass filter, etc.) can

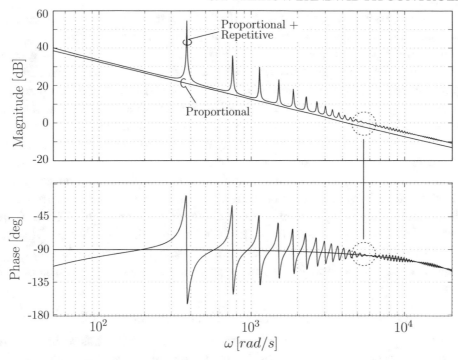

Figure 6.14: Open-loop system gain for the repetitive-based controller.

be useful for the generation of a continuous waveform, especially for higher M_c values. Indeed, the decimation rate can be even higher than what we have considered, since its limit is, theoretically, only represented by the Nyquist frequency for the highest-order harmonic one wants to compensate. Of course, practical issues related to system stabilization, i.e., its sensitivity to phase lag effects in the vicinity of the crossover frequency, actually compel to keep the decimation factor well below this theoretical limit.

The operation of the repetitive-based controller has been simulated with the UPS model already considered for testing the large bandwidth controllers. In order to better highlight the merits of this solution we have considered the typical situation where a distorting load, represented by a high crest factor diode rectifier with capacitive filter, is connected at the UPS output. Because of the non zero output impedance of the UPS, the load current peaks determine a typical distortion of the output voltage waveform. The repetitive controller is able to slowly compensate this distortion, reducing it to a minimum in a relatively large number of fundamental frequency periods. This is basically the situation depicted by Fig. 6.15. The figure was obtained applying, at first, only the proportional controller. The corresponding voltage distortion is shown in Fig. 6.15(b). After a few fundamental frequency periods, at instant $t = 0.1$ s, the repetitive-based controller is activated. Its operation generates a transient that extends through several fundamental frequency periods. That is due to the fact that, as the controller reduces the voltage distortion, the crest factor of the load current progressively increases. This typical regenerative effect, that is common to all uncontrolled rectifiers with capacitive filter, is described by the right column of Fig. 6.15, where the inverter output current and its reference are represented. In particular, comparing Fig. 6.15(c) with Fig. 6.15(e) and Fig. 6.15(d) with Fig. 6.15(f), it is possible to realize how the voltage waveform is corrected by the controller, and to appreciate the effect this causes on the load current. In the end, a new steady state is reached, where the voltage distortion is strongly attenuated, even if the load current crest factor has significantly increased.

As Fig. 6.15 clearly demonstrates, the performance of the repetitive-based controller can be quite satisfactory. Nevertheless, some caution is required in the implementation of this type of controller. Indeed, the settling time of the output voltage is in a the range of about 10 fundamental frequency periods. It is generally quite difficult to improve this significantly. This implies that, if more frequent load variations can be expected for the considered application, the controller effectiveness is likely to get lost, as it would be permanently operating in transient conditions.

6.4.1   THE DFT FILTER-BASED VOLTAGE CONTROLLER

A different interpretation of the repetitive control concept, that tends to improve some of its drawbacks while retaining the main positive features, is represented by what we call the *DFT filter-based* selective harmonic compensation strategy [14]. We are again referring to a narrow bandwidth controller, whose dynamic response is going to extend itself over several fundamental frequency periods. As the repetitive-based, also the DFT filter-based controller is conceived to operate in parallel with a conventional voltage regulator and to boost the loop gain only at certain, predefined frequencies of interest, that are normally some selected harmonics of the fundamental frequency. This concept is also closely related to that of the rotating reference controllers we have been considering in Chapter 5. Actually, the DFT filter-based controller can be considered an effective way to implement the same control strategy on multiple frequencies.

We have seen how the repetitive-based controller requires that the designer implements some filtering in the delay line to control the system phase in the vicinity of the crossover frequency. The choice of the filter and the control of its interaction with the delay line are the most difficult aspects of the repetitive controller design one has to tackle. The DFT filter-based approach tends to mitigate this problem.

The proposed controller organization can be seen in Fig. 6.16, where the two controller parallel components can be identified. The first is a rotating reference frame PI controller, that, as explained in Chapter 5, is fully equivalent to the structure of Fig. 6.16, where a resonant filter centered on the output voltage fundamental frequency is substituted to the integral part of the original PI controller. Please note that this equivalence holds even if the original system is single phase, since the rotating reference frame can be, as well, used to represent single-phase quantities [14]. From the implementation standpoint, however, once the equivalence is exploited and

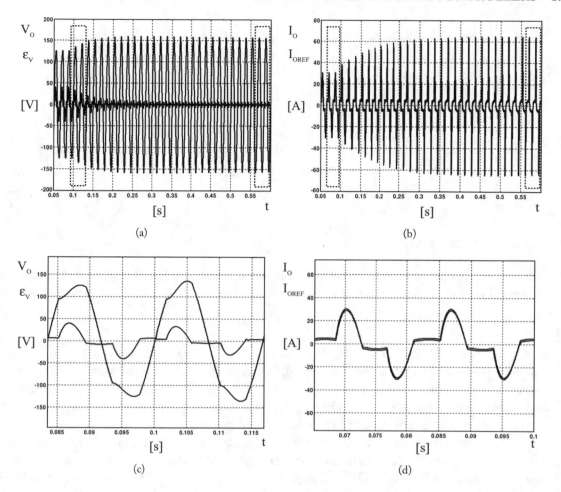

Figure 6.15: Repetitive-based controller operation. (a) Output voltage transient; (b) output current transient; (c) detail of (a) before the repetitive controller is activated; (d) detail of (b) before the repetitive controller is activated. *(Continues)*

the block diagram of Fig. 6.16 is derived, this interpretation of the rotating reference frame is no longer relevant. The rotating PI controller will guarantee zero steady-state tracking error on the fundamental component of the output voltage.

The second component of the considered voltage controller is designed to take care of high-order harmonics. As in the repetitive-based case, its function is to boost the system open-loop gain at certain predefined frequencies. To achieve this result, once again a positive feedback arrangement is considered. Of course, at any frequency where the gain of the $F_{DFT}(z)$ filter is unity and its phase is zero, the positive feedback will boost the controller gain to infinity. The nice thing

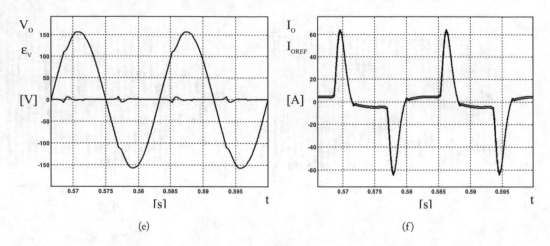

(e)   (f)

Figure 6.15: *(Continued)*. Repetitive-based controller operation. (e) Detail of (a) after the steady state is reached with the repetitive controller and (f) detail of (b) after the steady state is reached with the repetitive controller.

about this controller is that, by properly choosing the $F_{DFT}(z)$ filter, it is possible to have gain amplification *only* where it is actually needed, i.e., at pre-defined, selected harmonic frequencies, not at *each* harmonic frequency, as it happened for the repetitive-based solution. Please note that this allows to save the smoothing filter $F_1(z)$, whose design is typically quite complicated, and that was absolutely necessary for the repetitive-based controller.

To achieve the above-mentioned selective compensation and to get an adjustable phase lead, that may be required to ensure a suitable phase margin at the crossover frequency, we propose the use of "moving" or "running" DFT (Discrete Fourier Transforms) filters, with a window length equal to one fundamental period, such as

$$F_{DFT}(z) = \frac{2}{M} \sum_{i=0}^{M-1} \left(\sum_{h \in N_h} \cos\left[\frac{2\pi}{M} h\,(i + N_a)\right] \right) z^{-i}, \tag{6.22}$$

where N_h is the set of selected harmonic frequencies and N_a the number of leading steps required to get the phase lead that ensures system stability. Equation (6.22) can be seen as a Finite Impulse Response (FIR) pass-band filter with M taps, presenting unity gain at all selected harmonics in N_h. It is also called Discrete Cosine Transform (DCT) filter. One advantage of (6.22) is that the compensation of more harmonics does not increase the computational complexity, and the phase lead can be tuned at the design stage by parameter N_a. Of course, in order to implement the repetitive concept, a delay line with N_a taps is needed in the feedback path to recover zero phase shift of the loop gain $F_{DFT}(z)\,z^{-N_a}$, at the desired frequencies, which is the necessary condition

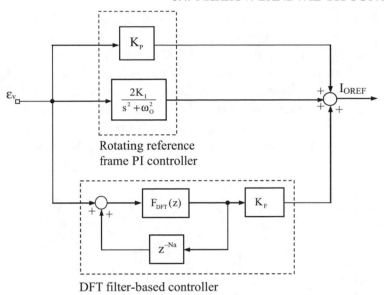

Figure 6.16: (a) Suggested DFT filter-based voltage controller. A rotating reference frame PI controller is parallel connected to the DFT filter-based controller.

to have gain amplification. Another advantage of (6.22) is that its structure is highly adapted to the typical DSP architecture, where the execution of multiply and accumulate instructions normally requires a single clock cycle. This makes the DFT-based controller extremely effective, in particular if compared to the implementation of a bank of resonant filters.

Considering now our example case, we would like to briefly outline the design procedure for the DFT filter-based voltage controller. The rotating reference frame PI design is straightforward: a conventional digital PI is designed for the UPS (Section 6.3.1, Aside 12) and then turned into the rotating equivalent of Fig. 6.16. This requires simply the doubling of the proportional and integral gains for the resonant filter part of the regulator (as we did in Section 5.3.2 for the PI current controller).

The design of the DCT filter is quite easy as well: since we don't need to recover the system phase, thanks to the relatively high ratio between sampling frequency and required crossover frequency, parameter N_a can be simply set to zero. The number of filter taps is then given by the ratio between the sampling frequency and the fundamental output voltage frequency, that, in order to avoid leakage effects on the DFT filter, must be an integer number. Because of this constraint, as we did before, we slightly changed the sampling frequency to 19.8 kHz so as to get $M = 330$. The Bode plot of the obtained open-loop gain is shown in Fig. 6.17. It is interesting to compare this figure to Fig. 6.14. As can be seen, gain amplification takes places only at the predefined frequencies, determining little effects on the system-phase margin. The stability of the closed-loop system, is consequently determined by the PI controller's design, as in a conventional implementation. In order to limit the computational effort and the memory occupation a sample decimation by a factor M_c can be used in the FIR filter implementation, similarly to what we have done for the repetitive control. More precisely, in our example, M has been reduced by a factor 10 ($M_c = 10$, $M = 33$) without significantly affecting the dynamic performance. Similarly, to the repetitive control, the main issue related to the use of decimation is that the output of the DCT filter is updated only every M_c samples and it is seen by the proportional controller as a stepwise function. In order to emulate an interpolator, a moving average filter with M_c taps has been adopted.

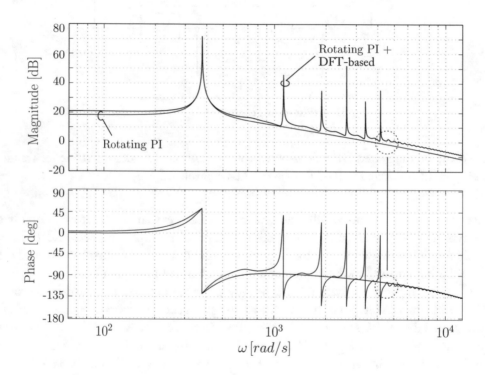

Figure 6.17: Open-loop system gain for the DFT-based controller.

As far as the design of the gain K_F is concerned, we can follow the same guidelines that we have illustrated in the Aside 10, when we described the design of a stationary reference current controller with zero steady state error. This may seem surprising, at first, but we must recall that the DFT filter is nothing but a bank of parallel resonant filters, each tuned on one of the harmonics to be compensated. In Chapter 5, we have exactly shown that a rotating reference controller is also equivalent to a tuned resonant filter, therefore, the same criteria can be adopted for the design of the controller gain in both cases [14]. In the end, the effect of this gain is to determine the settling time of the DFT-based controller to any disturbance. In the considered example, it was set to a value corresponding to a settling time equal to 10 fundamental periods.

To complete the design, we still need to specify the set of harmonics we want to compensate, N_h. In our example case, this was set to $\{3, 5, 7, 9, 11\}$.

The controller operation is illustrated by Fig. 6.18, that considers the UPS system behavior in the same conditions of Fig. 6.15. Once again, the controller initially operates only in PI mode. This implies a significant output voltage distortion, that can be observed in Fig. 6.18(c). After 0.1 s, the DFT filter-based section of the controller is activated, determining the progressive attenuation of the voltage tracking error. As in the previous case, the interaction between the UPS output impedance and the diode rectifier determines an increase in the load current crest factor, as can be seen comparing Fig. 6.18(d) and Fig. 6.18(f). An important difference with the previous example is represented by the internal current controller: in this case a purely proportional current regulator was employed. This is the reason why the current tracking error, visible on the left column of Fig. 6.18, is somewhat higher than that we can observe in Fig. 6.15. Nevertheless, considering the right column of Fig. 6.18, we can appreciate the very satisfactory performance of the DFT-based controller. This allows us to conclude that, as far as a narrow bandwidth voltage controller's effectiveness is concerned, the presence of a high-performance internal current controller is not essential. Indeed, in the steady state, the quality of the harmonic compensation can be very high. Of course, in dynamic conditions, i.e., in the presence of load step changes or other fast transients, the system's speed of response and its damping, that are functions also of the current loop bandwidth, could be unacceptable. However, in case a limited bandwidth current controller has to be accepted, the phase lead effect of the DFT controller can be exploited to increase the system's phase margin and push the bandwidth very close to the limit.

6.5 OTHER APPLICATIONS OF THE CURRENT CONTROLLED VSI

We would like to conclude the discussion of external control loops for current controlled voltage source inverters by briefly describing a couple of other important applications, where the multi-loop organization is often employed. These are the controlled rectifier and the active power filter.

They are fundamentally similar, the hardware organization being exactly the same. In both applications the VSI is connected to a primary source of energy, that can be simply the utility grid or any other, more complex, power system. In both of them, the VSI has to impose a predefined, controlled current onto the source. The main difference between the two is represented by the fact the controlled rectifier directly supplies power to a DC load, while the active power filter not necessarily does, being typically employed only to compensate undesired harmonic current components and/or reactive power injected into the source by other distorting and/or reactive loads. Because of that, the design criteria adopted for the power converter can be different in the two cases. In order to visualize the typical organization of both these applications we can refer to Fig. 6.19.

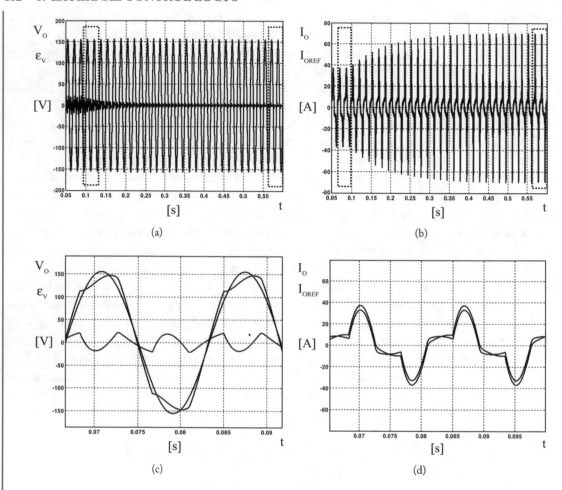

Figure 6.18: DFT filter-based controller operation. (a) Output voltage transient; (b) output current transient; (c) detail of (a) before the repetitive controller is activated; and (d) detail of (b) before the repetitive controller is activated. *(Continues)*

As can be seen, the VSI, that can be single or three phase, is normally connected to the AC power source through an input filter. This is used to attenuate the high frequency components of the converter output current injected into the source. Apart from that, we can immediately recognize the same basic structure we have been considering in our discussion of current control implementations. We can therefore conclude that, with the exception of minor modifications that may be required to take the input filters into account, current controllers for PWM rectifiers and active filters can be based exactly on the same concepts we have been considering in previous chapters. Although it is possible, at least from the general organization point of view, to treat

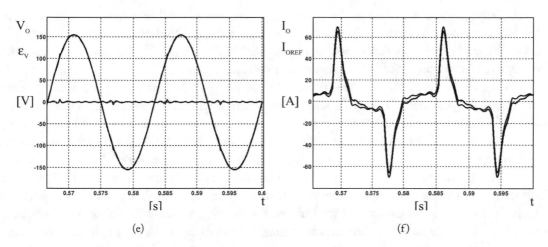

(e)   (f)

Figure 6.18: *(Continued)*. DFT filter-based controller operation. (e) Detail of (a) after the steady state is reached with the DFT-based controller and (f) detail of (b) after the steady state is reached with the DFT-based controller.

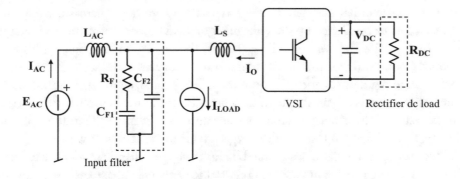

Figure 6.19: Typical organization of a controlled rectifier or active power filter.

the two applications in a unified manner, the different goals of the rectifier and the active filter sometimes call for different control strategies. Therefore, we will now analyze them separately.

6.5.1   THE CONTROLLED RECTIFIER

The PWM controlled rectifier can be represented by Fig. 6.19 once the I_{LOAD} generator is not considered and an equivalent DC load, represented for simplicity by resistor R_{DC}, is connected to the converter output. The typical control objective for this converter is to supply the load with controlled DC power, absorbing high quality (i.e., harmonic and reactive component free) AC power from the source. This requires two different control loops: (i) a current control loop, that is used to impose an AC current I_{AC} on the source, proportional to the input voltage E_{AC} and (ii) a DC voltage control loop, that is used to regulate the load voltage, V_{DC}, keeping it to a predefined value, even in the presence of load and/or line voltage variations.

The current control loop does not need to be particularly fast: indeed the typical reference waveform, proportional to the AC source voltage, is represented by a practically sinusoidal signal. Even if the source were affected by a significant harmonic distortion, a current loop bandwidth in the order of 10–20 times the source fundamental frequency would allow to track the reference without appreciable errors. This is the typical grounds for the application of PI current controllers.

In the case of a three-phase power system, the modulator and current controller can take advantage of the techniques we have been discussing in Chapter 5. These become particularly useful in case we are considering a medium power rectifier, rated for several tens of kVA. In that case, it is likely that the sampling and switching frequency has to be kept relatively low, making it difficult to guarantee a good reference tracking even at the fundamental frequency. Rotating reference controllers, possibly implemented as banks of resonant filters, are in this case particularly effective.

As far as the outer control loop is concerned, its goal is to adjust the current reference amplitude so as to keep the load voltage on the desired set-point. In single-phase systems, the instantaneous power unbalance determines a typical voltage ripple across the DC link capacitor [1, 2], that has to be filtered by the voltage regulator in order not to determine input current distortion. This implies the need for a limitation of the regulation loop bandwidth to a fraction, typically about one tenth, of the fundamental input frequency. Because of that, the design of the output voltage regulator is normally quite easy, being the dynamic specifications not so stringent. Once again, this is a typical situation where a PI controller is probably the best choice. In three-phase systems, the input power is constant and there is not any instantaneous unbalance. Nevertheless, the voltage loop bandwidth is again typically relatively low.

To design the PI regulator, a suitable DC link voltage dynamic model has to be derived. In order to sketch a design procedure, that is referred to the single phase case, we must first realize that the voltage controller actually determines the amount of power absorbed by the rectifier from the AC source. In the steady state, this has to be equal to the sum of the load power and the converter losses. Instead, in dynamic conditions, the DC link capacitor can absorb or deliver the instantaneous power unbalance. Therefore, the fundamental equation that describes the power balance of the system is the following:

$$\frac{d}{dt} E_{C_{DC}} = P_{AC} - P_{loss} - P_{LOAD}. \tag{6.23}$$

In (6.23), $E_{C_{DC}} = \frac{1}{2}C_{DC}V_{DC}^2$ is the energy stored in the DC link capacitor, P_{loss} is the power the converter dissipates, $P_{LOAD} = V_{DC}^2/R_{DC}$ is the power delivered to the load, and P_{AC} the input active power, under the hypothesis of unity power factor rectifier operation, is given by

$$P_{AC} = G_{EQ} \cdot E_{AC_{RMS}}^2, \tag{6.24}$$

where G_{EQ} represents the voltage controller output. This, as stated above, represents the desired amplitude of the source current, whose waveform is assumed to be proportional to that of the source input voltage E_{AC}. Rewriting (6.23) in terms of the system parameters we find the following dynamic equation:

$$\frac{1}{2}C_{DC}\frac{d}{dt}V_{DC}^2 = G_{EQ}E_{AC_{RMS}}^2 - P_{loss} - \frac{V_{DC}^2}{R_{DC}}, \tag{6.25}$$

which relates the controller variable, V_{DC}, with the controller's output G_{EQ}. As can be seen, (6.25) is a nonlinear differential equation; therefore, before a dynamic model can be derived, a linearization procedure has to be applied. Of course, since the linearization is based on variable perturbation and small-signal approximation, the model will be only valid in the vicinity of a steady state operating point. However, it is interesting to note that, if V_{DC}^2 is chosen as the controlled variable, (6.25) becomes linear and can be directly used for the derivation of the system dynamic model, that, in this case, will be valid also for large signals. In other words, controlling V_{DC}^2 instead of V_{DC}, which is functionally equivalent, can greatly extend the linearity of the control loop.

In practice, since the DC link voltage V_{DC} is almost constant, being only affected by a relatively small low frequency ripple, the difference in the achievable performance between the two possible approaches is very small.

Linearization of (6.24), is done assuming that $E_{AC_{RMS}}$ and P_{loss} are constant and considering, as usual, each variable quantity to be equal to the superposition of a steady-state component and a perturbation component, i.e., $V_{DC} = \overline{V}_{DC} + v_{dc}$, $G_{EQ} = \overline{G}_{EQ} + g_{eq}$ with obvious meaning of the symbols. Simple calculations yield the following result:

$$\frac{v_{dc}}{g_{eq}}(s) = \frac{R_{DC}E_{AC_{RMS}}^2}{2\overline{V}_{DC}}\frac{1}{1 + sC_{DC}\frac{R_{DC}}{2}}, \tag{6.26}$$

that can be used in the design of the DC link voltage regulator. The design of the regulator can follow the same steps we followed in Chapters 2 and 3, with continuous time synthesis and successive discretization. The only caution we need to take is to limit the required bandwidth and keep it lower than the source fundamental frequency, so as to avoid source current distortion.

6.5.2 THE ACTIVE POWER FILTER

The active power filter application can be represented by Fig. 6.19 as well. In this case, the I_{LOAD} generator is considered and used to represent the distorting or reactive loads the filter has to compensate, while the DC load, R_{DC}, might not be present. If there is no DC load, the active power filter is not required to process any active power, with the exception of that due to its losses, and can thus be sized to sustain only the reactive and harmonic load currents. A typical control objective for this application is to compensate the harmonic and reactive load currents, so as to make the AC source current proportional to the source voltage. This implies that, from the source standpoint, the load will be seen as an equivalent resistor, absorbing only the active power required by the distorting loads. The achievement of this objective requires again two different control loops: (i) a current control loop, used to impose the desired AC current I_{AC} to the source and (ii) a DC voltage control loop, that is used to regulate the load voltage, V_{DC}, keeping it equal to a given reference value.

Apparently, this situation seems identical to that of the rectifier discussed in the previous paragraph. This is actually the case for the voltage loop, that can be designed exactly as that of the rectifier. It is not at all the case for the current loop: the compensation of high order harmonic currents normally require some high-performance current control loop. Indeed, the implementation of a simple PI current controller is normally able to offer only a limited harmonic compensation capability, that is very often quite far from being satisfactory.

Therefore, more complex solutions have to be taken into account. As we have illustrated for the UPS voltage loop, in this case it is as well possible to follow two different design philosophies: *(i)* implementing a large bandwidth current controller or *(ii)* implementing a narrow bandwidth current controller. The former solution is aimed at the instantaneous compensation of any deviation of the current injected into the line from its reference waveform. The latter is instead aimed at the slow compensation of the same deviation, typically requiring several fundamental frequency periods to be accomplished.

The large bandwidth controllers, that, in the digital domain, are exactly of the predictive type we have been discussing in Chapter 3 and Chapter 4, are normally suited to all those situations where the distorting and harmonic load currents are characterized by unpredictable and frequent variations. In some particularly demanding cases, hysteresis controllers can be used as well.

The narrow bandwidth controllers can be based on the resonant filters or, equivalently, on the rotating reference frame regulators seen in Chapter 5. In the active filter application, several parallel regulators will be implemented to take care of the different harmonic frequencies to be compensated. Repetitive- or DFT-filter based controllers, of the type seen in Section 6.5, are also viable solutions. Of course, since the dynamic response of these regulators normally extends on some fundamental frequency periods, their adoption should be limited to those cases where the distorting and reactive load currents are not subject to frequent variations and therefore the controller steady state is not too frequently perturbed. The design of the narrow bandwidth regulators exactly follows the principles we have illustrated for the UPS voltage control case.

The last issue we need to examine to complete this brief description of active power filter control, is related to the generation of the inverter reference current signal. From Fig. 6.19 we can see that, in order to achieve the desired compensation and inject a voltage proportional current into the AC source, the inverter simply needs to generate a current equal to the algebraic sum of the desired source current and the load current. Therefore, in the most straightforward approach, the inverter current reference can be built as:

$$I_{OREF} = -I_{AC}^* + I_{LOAD} = -G_{EQ}E_{AC} + I_{LOAD}, \qquad (6.27)$$

where G_{EQ}, as in the rectifier case, represents the output of the DC link voltage regulator. Of course, the implementation of (6.27), is relatively simple only if the measurement of the distorting loads' current I_{LOAD} is possible. If this is the case, the result of its application will be the cancellation of the reactive current component from the AC source current. In addition, any harmonic current not present in the AC source voltage will also be canceled. The quality of the cancellation is, of course, limited only by the chosen current controller reference tracking capabilities [15].

If current I_{LOAD} cannot be measured, or if the active power filter is designed for more complex tasks, like the partial, controlled compensation of some selected harmonics and/or the compensation of the load reactive power only, different approaches for the computation of the converter current reference can be employed, whose illustration, however, goes beyond the scope of this book.

REFERENCES

[1] N. Mohan, T. Undeland, W. Robbins, "Power Electronics: Converters, Applications and Design," 2003, Wiley, ISBN 0–471-22693-9. 151, 184

[2] J. Kassakian, G. Verghese, M. Schlecht, "Principles of Power Electronics," 1991, Addison Wesley, ISBN 0–201-09689-7. 151, 184

[3] Y. Dote and R.G. Hoft, "Intelligent Control–Power Electronic Systems," Oxford University Press Inc., 1998. DOI: 10.1109/MPER.1999.785805. 151

[4] M.J. Tyan, W.E. Brumsickle, R.D. Lorenz, "Control Topology Options for Single-Phase UPS Inverters," *IEEE Transactions on Industry Applications,* Vol. 33, No. 2, March/April 1997, pp. 493–500. DOI: 10.1109/MPER.1999.785805. 151

[5] S. Buso, S. Fasolo, P. Mattavelli: "Uninterruptible Power Supply Multi-Loop Control Employing Digital Predictive Voltage and Current Regulators," *IEEE Transactions on Industry Applications,* Vol. 37, No. 6, November/December 2001, pp. 1846–1854. DOI: 10.1109/28.968200. 161, 169

[6] O. Kükrer, "Deadbeat Control of a Three-Phase Inverter with an Output LC filter," *IEEE Transactions on Power Electronics,* Vol. 11, No. 1, January 1996, pp. 16–23. DOI: 10.1109/63.484412. 164, 169

[7] O. Kükrer, H. Komurcugil "Deadbeat Control method for single-phase UPS inverters with compensation of computational delay," *IEE Proceedings – Electric Power Applications,* Vol. 146, No. 1, January 1999, pp. 123–128. DOI: 10.1049/ip-epa:19990215. 164, 169

[8] A. Kavamura, T. Haneyoshi, R.G. Hoft, "Deadbeat Controlled PWM Inverter with Parameter Estimation Using only Voltage Sensor," *IEEE Transactions on Power Electronics,* Vol. 3, No. 2, April 1988, pp. 118–124. DOI: 10.1109/63.4341. 164, 169

[9] T. Yokoyama, A. Kawamura, "Disturbance Observer Based Fully Digital Controlled PWM Inverter for CVCF Operation," *IEEE Transactions on Power Electronics*, Vol. 9, No. 5, September 1994, pp. 473–480. DOI: 10.1109/63.321031. 169

[10] P. Mattavelli, "An improved Dead-beat Control for UPS using Disturbance Observers," *IEEE Transactions on Industrial Electronics*, Vol. 52, No. 1, February 2005, pp. 206–212. DOI: 10.1109/TIE.2004.837912. 169

[11] M. Morari, E. Zafiriou, "Robust Process Control," January 1989, Prentice Hall, ISBN 0-137-82153-0. 170

[12] Y.Y. Tzou, R.S. Ou, S.L. Jung, M.Y. Chang, "High-Performance Programmable AC Power Source with Low Harmonic Distortion Using DSP-Based Repetitive Control Technique," *IEEE Transactions on Power Electronics*, Vol. 12, No. 4, July 1997, pp. 715–725. DOI: 10.1109/63.602567. 170, 171

[13] K. Zhang, Y. Kang, J. Xiong, J.Chen, "Direct Repetitive Control of SPWM Inverters for UPS Purpose," *IEEE Transactions on Power Electronics*, Vol. 18, No. 3, May 2003, pp. 784–792. DOI: 10.1109/TPEL.2003.810846. 170

[14] P. Mattavelli, "Synchronous frame harmonic control for high-performance AC power Supplies," *IEEE Transactions on Industry Applications*, Vol. 37, No. 3, May/June 2001, pp. 864–872. DOI: 10.1109/28.924769. 176, 180

[15] S. Buso, L. Malesani, P. Mattavelli: "Comparison of Current Control Techniques for Active Filter Applications," *IEEE Transactions on Industrial Electronics*, Vol. 45, No. 5, October 1998, pp. 722–729. DOI: 10.1109/41.720328. 187

CHAPTER 7

New digital control paradigms

In the presentation of the digital current controllers of Chapter 3, we have assumed their implementation to be based on *software programmable* devices, like microcontrollers or digital signal processors. Indeed, this is, by far, the most commonly adopted approach to digital control in power electronics.

As we already mentioned, however, different approaches can be followed, that, in our opinion, are going to become the standard practice in the future. In this chapter, we would like to discuss the possible alternatives to the *software* approach and the related application fields, with a particular emphasis on what we will define as *complex control architectures*.

7.1   FLEXIBILITY VS. PERFORMANCE

In comparison with software-based solutions, custom designed digital control chips represent a radically different implementation strategy for digital controllers. Their usage is basically aimed at one or both of the following objectives: (i) optimizing the cost of the controller, tailoring the hardware to the exact functionality it is called to provide; and (ii) maximizing the performance level of a given power converter topology, by compressing computation times to the minimum and pushing switching frequency, small-signal control bandwidth and large-signal speed of response as high as possible. Following this approach, some leading integrated circuit manufacturers are nowadays offering application specific digital control chips, designed and optimized for particular DC-DC converter topologies and applications, such as, for instance, [1] or [2].

Software-based control solutions offer, inherently, maximum flexibility and adaptability, but, as we have seen, their performance is limited by the AD converter latency and the computation delays. The custom designed hardware solutions, instead, achieve very high performance levels, but offer little or no flexibility. The lack of flexibility is certainly one of the factors that are making fully integrated digital controllers not so successful in taking the place of analog control ICs, whose usage is still very large. Especially for large-scale productions, the cost of analog controllers is very hard to beat, and not always the additional features a digital controller can offer, with respect to an analog one, can pay back the increase in the cost.

Comparing the two digital control approaches, i.e., the software-based and custom hardware based ones, a clear trade-off between flexibility and performance appears, that, apparently, is difficult to avoid.

7.2 FLEXIBILITY AND PERFORMANCE

Carefully observing the digital control device portfolios of leading manufacturers, it is possible to recognize that a novel design approach is becoming more and more affordable, one that is based on *hardware programmable* logic devices, in the form of Field Programmable Gate Array (FPGA) chips, and large bandwidth, low-latency AD converters.

Indeed, the cost of FPGA chips is steadily decreasing, although the gate density, i.e., the number of combinatorial and sequential logic blocks that can be fitted into a single device, is increasing at impressive rates. As a result, a larger and larger number of reasonably cheap and extremely high performing programmable digital circuits is currently available on the market. Some of them, like [3], are actually hardware *and* software programmable devices, where, on the same chip, one or more microcontrollers, FPGAs and even analog amplifier circuits are allowed to operate concurrently. These represent what can be called a complex, mixed-signal *system on chip* (SoC). Other manufacturers are following this trend, proposing FPGA systems so large to include a fully functional microprocessor, that can be hardware-synthesized at need, while leaving a large amount of programmable gates available for the user application.

At the same time, large bandwidth, low-latency, pipeline-based AD converters are becoming, likewise, more and more economical, allowing control designers to implement multi-mega-sample-per-second (M*Sample*/s) data paths at very reasonable costs. Sometimes, these devices are integral parts of the above-mentioned SoCs, that therefore represent ready to use, fully functional control devices.

To facilitate the use of these technologies with a more conventional approach, ready to use *controller boards* like, e.g., [4, 5], are also becoming relatively popular. They offer a combination of software programmable devices, FPGA chips and AD converters that are integrated on the same board. Thanks to the presence of high speed communication buses connecting the different component integrated circuits, of a basic pre-installed firmware configuration and of a flexible analog signal conditioning circuitry, they are relatively easy to program and rapidly adapt to different user applications. From these standpoints, they represent almost ideal rapid digital control prototyping platforms.

Altogether, the mixed signal SoCs and control boards are certainly going to determine a thorough revision of digital controller architectures, allowing to achieve, *at the same time*, flexibility *and* performance. What we can call a new power converter digital control *paradigm* is, indeed, on its way to become the standard practice. Thanks to high sampling frequencies and to programmable high-speed computing hardware, it will enable the design of a whole lot of innovative digital controllers.

The theoretical foundations of digital power converter control are, of course, totally independent from the chosen implementation platform or device. On the other hand, the use of modern control hardware platforms enables the realization of architectures that are, in many ways, outside the reach of conventional software- and/or firmware-based controllers.

As a result, when discussing these new technologies, we do not need to deepen our knowledge of the basic control techniques we have already examined in the previous chapters, because these will remain essentially unchanged, but rather to extend our vision towards higher levels of abstraction, where the full potential of high performance control platforms can be exploited.

As a simple example of that new paradigm, we have discussed in Chapter 4 multi-sampled current controllers and seen how they represent a viable option to significantly improve a current controlled converter's performance. But even more challenging and complex applications can actually take advantage of it, one of which is going to be described in the following section.

7.3 DISTRIBUTED GENERATION CONTROL ARCHITECTURES

The multi-layer control architectures required by distributed electrical power generation and renewable energy exploitation [6, 7, 8, 9, 10] represent good examples of higher complexity digital control applications. They also represent one of the ultimate application fields for all the techniques we have discussed throughout this book and, in our opinion, certainly deserve a closer look. Therefore, in the rest this chapter, we would like to address the issues related, precisely, to the implementation of the so called *smart micro-grid* controllers.

With the term smart micro-grid, we refer to those low-voltage electrical energy distribution grids where multiple energy generators and energy storage devices operate concurrently. The micro-grids can be connected to the utility distribution grid at a single point, denoted as Point of Common Coupling (PCC) or even operate autonomously, in islanded mode. Both the generators and the storage devices that populate the micro-grid have to be controlled individually, but also coordinated with one another, so as to maintain, in all possible operating conditions, the power quality experienced by the end-users (i.e., the loads) at the desired level, with, e.g., nominal voltage amplitude and frequency, minimum voltage harmonic distortion, minimum harmonic current injection at the PCC, minimum distribution losses, etc.

Accordingly, the smart-grids host a relatively large number of converters, voltage source inverters of the same type of our case study or possibly with a full bridge topology, that are used as Distributed Energy Resource (DER) or electrical Energy Storage (ES) interface converters. They basically govern the interaction among the energy sources, the storage devices, the micro-grid and the loads. To this purpose, they are typically required to operate in quite sophisticated ways, not just as controlled voltage or current sources, but rather as controlled *active and/or reactive power sources*. All converters have to operate coordinately, either as the result of high level supervision and point to point communication, or thanks to distributed, multi-agent network control algorithms, or else by the adoption of robust and automatic power sharing control strategies.

All these solutions call for a particular control system organization, where multiple functions, at different hierarchical levels, need to be concurrently performed. The distinctive feature that makes these *complex control architectures* quite different from conventional multi-loop or multi-variable control organizations is their being distributed among a number of processing circuits. The low-level functions are executed locally, while middle level ones are *shared* among different controllers and, therefore, require communication channels to achieve a coordinated operation through data exchange. The highest level ones are often hosted by remote computers, that interact with the distributed converter controllers at relatively low frequency, for monitoring and long term optimization purposes. We will now see how these quite sophisticated control architectures can be more effectively implemented with high performance and flexible platforms rather than with any other conventional device, like a DSP or microcontroller board.

7.3.1   VSIS IN DISTRIBUTED GENERATION GRIDS

The typical configuration of a DER/ES interface converter, suitable to be used in a single-phase smart micro-grid, is shown in Fig. 7.1. In particular, Fig. 7.1(a) shows the overall converter block diagram, comprising a DC-DC stage, that deals with the DER primary energy source (e.g., a photovoltaic generator, a fuel cell) or energy storage device, and a DC-AC stage, that takes care of grid interfacing. From now on, we will refer only to the *second* stage, whose basic electrical scheme is shown in Fig. 7.1(b). The converter has a full bridge topology, but a half bridge inverter topology can be equivalently applied. The full bridge is somewhat more frequently encountered in commercial products, thanks to its advantages in terms of switch voltage stress.

As can be seen, the two conversion stages typically share a single control board. As far as the inverter stage is concerned, the main purpose of the controller is to enable locally synchronized current injection/absorption, so as to: (i) guarantee the maximum exploitation of the energy source or the appropriate utilization of the storage element (which impacts on the delivered/absorbed real power, P) and (ii) concur to the determination of a satisfactory power quality throughout the grid (which may require controlled reactive power, Q, injection/absorption). While the former objective can be met using locally available data, the latter normally requires interaction with other inverters, in a distributed control scenario, or with a central supervising unit, in a centralized control scenario.

7.3.2   CONTROL SYSTEM ORGANIZATION

In general terms, the controller's functions for an application like that of Fig. 7.1 can be represented as a stack, i.e., a multi-layer hierarchical architecture, where lower level functions receive inputs from upper level ones. In turn, they provide data to the upper level functions that are used in the decision making algorithms, governing, e.g., the power exchange with the micro-grid.

The typical structure of the control stack is shown in Table 7.1. As can be seen, the lower level functions, namely PWM, synchronization, electrical variable controllers, are supposed to be implemented on a programmable logic device, like a FPGA. The basic reason is that, as we have seen, they do not involve elaborate signal processing (fixed point arithmetic is perfectly viable), but require relatively high cycle frequencies, in the tens of kHz range. In addition, they require internal, jitter-free synchronization, for example between sampling and modulation processes,

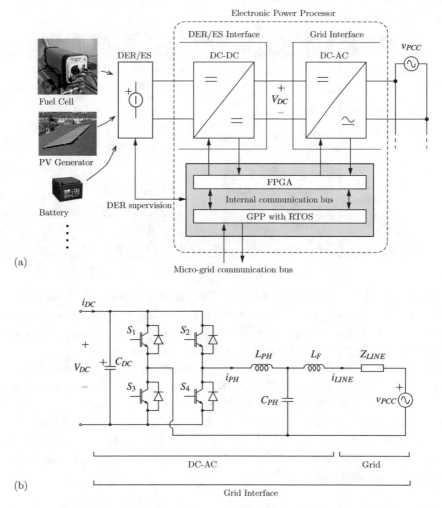

Figure 7.1: Typical DER coupling converter configuration: (a) block diagram illustrating the different conversion stages and related control signals and (b) basic schematic diagram of the DC-AC stage.

and reliable fault detection mechanisms. These features are easily implemented in a configurable digital chip.

Instead, higher level functions, such as P and Q control or local grid characterization and optimization algorithms, make extensive use of signal processing functions, where floating point arithmetic can be very advantageous, and, especially in the case of distributed implementations, of advanced communication capabilities.

On the other hand, they do not necessarily require tight synchronization or strictly constant cycle frequencies. As a result, the natural implementation hardware for these functions is a

Table 7.1: Control function stack

Subsystem	Functionality	Scope	Target
Inverter	PWM	Local	FPGA
	PLL	Local	
	Current loop	Local	
	Fault detection	Local	
	P loop	Local	RT GPP
	Q loop	Local/Distributed	
	Grid voltage control	Local/Distributed	
	Grid characterization	Local/Distributed	
	Global optimization	Distributed	
DC/DC	PWM	Local	FPGA
	Input V/I control	Local	
	Source/storage optimization	Local/Remote	RT GPP

DSP or, more and more frequently, a General Purpose Processor (GPP). Both devices are programmable in high level languages, which increases abstraction and flexibility. The GPP normally has the advantage of supporting a Real-Time Operating System (RTOS), a very useful resource for this application, where reliability and determinism are mandatory.

An exhaustive discussion on digital devices suitable for the considered power electronics control application is proposed in [11], but our conclusion is that the best-suited controller for a micro-grid DER coupling converter is a multi-platform digital device, comprising, at least, two domains: (i) a high-performance hardware programmable device (e.g., a FPGA) and (ii) a DSP or GPP, possibly supporting a RTOS. Therefore, the DER converter controller we are now going to illustrate has been conceived to be deployed on one of such control devices, namely [5], whose basic features are summarized in Table 7.2.

7.3.3   CONTROL ARCHITECTURE IMPLEMENTATION

A fully functional example of the control stack for the inverter stage of a DER coupling converter like the one shown in Fig. 7.1 can be developed according to the scheme of Fig. 7.2. The different stacked control functions are partitioned referring to the hardware resources available on the chosen control board, so that, as can be seen, the controller is partly deployed on the FPGA module, partly on a RT GPP. The RT GPP hosts the higher level control functions and manages data logging and Ethernet communication with the centralized grid supervisor, a desktop PC in our set-up.

At the bottom level of the control stack, symmetrical, triangular digital pulse width modulation is implemented, with an equivalent 12.4 bit resolution for the duty-cycle, as per (2.4). On top of that, a proportional-integral current controller with 2 kHz bandwidth and $60°$ phase margin has been applied. Its design follows the same guideline we have presented in Section 3.2. The

Table 7.2: Control platform characteristics (GPIC)

Feature	Parameter	Value
Processor	Model Processor Speed	PowerPC 400 MHz
Memory	Nonvolatile System	512 MB 256 MB
FPGA	Model # Slices # DSP48s	Xilinx Spartan-6 LX45 6822 58
Network	Network interface	IEEE 802.3 Ethernet
Communication	Port	RS-232, RS-485 CAN, USB
Peripherals	Channel	16 AI, 12-bit, ±10 V, 100 kHz 14 ch., 500 kHz gate drivers [1]

basic controller parameters, together with the considered voltage source inverter characteristics, are listed in Table 7.3.

The reference for the current controller comes from a d-q rotating frame, that allows to easily manage the active and reactive current components. Differently from what we have seen in Chapter 5, the d-q transformation is applied to a single phase system, which requires the construction of a "virtual" α axis current component. This is done exploiting the simple Phase Locked Loop (PLL) circuit visible at the bottom of Fig. 7.2. The circuit creates a 90° lagging replica of the measured grid voltage signal, basically delaying the input by a quarter of its fundamental period. Under the assumption of limited voltage distortion, the application of (5.16) and the feedback arrangement allows the circuit to lock onto the input signal fundamental frequency, f_0, and to automatically track it when necessary. Besides, it determines the instantaneous phase angle θ, required to define the active and reactive reference current components, and the grid voltage fundamental component peak amplitude, V_{GRID}. In order to ensure a stable dynamic response to the phase locked loop, a local PI regulator needs to be suitably designed, as discussed, for instance, in [12]. Basically, its cross-over frequency needs to be limited below the grid voltage fundamental frequency.

Different controller organizations are actually possible, such as one where the current controller is itself implemented in the d-q rotating reference frame, which, as we know, guarantees an excellent fundamental frequency tracking capability, or others, thoroughly different, where the inverter is controlled in voltage mode, like in a UPS application. Indeed, the type of local control functions required to the inverter can be different, depending on the chosen micro-grid control methodology.

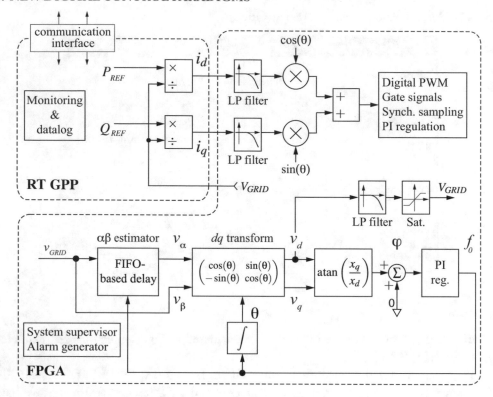

Figure 7.2: Controller implementation scheme on a control board of the type described by [5]: the controller is partly deployed on the FPGA module, partly on the RT GPP available on board.

Because we want to keep our presentation as simple as possible, we are not going to enter into the details of the high level micro-grid optimization and supervision functions that can be hosted by the GPP, because a whole lot of solutions are currently being investigated. Actually, the study of micro-grids and of the related control techniques is still far from being concluded and a widely recognized, effective control strategy has still to be found. Therefore, the solution presented in Fig. 7.2 has to be considered just a very simple example of what can be requested to the control system of a DER interface converter.

Accordingly, the RT GPP is just used to calculate the current reference components from the desired active, P_{REF}, and reactive, Q_{REF}, power set-points, received through the communication channel, and the local grid voltage amplitude, determined by the PLL.

It is important to underline that this simplicity is actually not necessary. Indeed, thanks to the control hardware organization, many other functions could be run on the GPP *without interfering* with the low-level control functions. The latter, being run by the FPGA sub-system, can proceed unperturbed in the background, irrespective of the complexity of the software running on the GPP.

Table 7.3: Inverter and PI current controller parameters

Parameter	Symbol	Value	
Nominal input voltage	V_{DC}	400	V
Switching frequency	f_s	18	kHz
Filter inductance	L_{PH}	1.2	mH
Series inductor resistance	$r_{s,PH}$	50	mΩ
Output filter capacitance	C_{PH}	10	μF
Output filter inductance	L_F	45	μH
Current sense gain	G_{TI}	375	mV/A
Voltage sense gain	G_{TE}	17.25	mV/V
Nominal power	S_o	3	kVA
Crossover frequency	f_{CR}	2	kHz
Phase margin	Φ_M	60	°
Carrier amplitude	$\pm A_r$	± 2777	
A/D conversion delay	Δt_{AD}	10	μs
PI calculation delay	Δt_{calc}	0.1	μs

Aside 14. Key Issues in Multi-Platform Control Architecture Implementation

The implementation of the control stack described by Table 7.1 on a multi-platform digital controller poses several technical questions [13], the main ones being related to:

1. analog to digital conversion;

2. data consistency between platforms;

3. development tools;

4. debugging.

AD Conversion

Using FPGAs for low level inverter control normally requires the design of additional hardware including the ADC chip, signal conditioning, and data bus synchronization circuitry. A possible solution for this problem is discussed in Chapter 4, referring to the multi-sampled current controllers.

In general, the custom design of high throughput data paths complicates significantly the controller development. However, in some systems, like the one described in [3] or [5], the ADC unit, and a lot of other I/O functions, are natively interfaced to the FPGA, so that the user does not need to implement the data path between the different devices.

For SoC architectures, this comes with practically no performance limitation (i.e., peripheral units determine the maximum data throughput), while in multi-chip boards a certain degree of latency is introduced, which typically limits the maximum data acquisition frequency to about 10 Mbit/s. This represents no real limitation though, as the bandwidths required to the current control loops in inverter applications are often in the kHz range and, as such, are perfectly compatible with these AD operation frequencies.

Data Consistency

Perhaps the most complex issue in using multi-platform controllers, however, is guaranteeing data consistency among the different levels of the control function stack.

The basic approach, where the logic circuitry is treated as a memory mapped external peripheral unit, while data exchange and task scheduling in the DSP, or GPP, is regulated by interrupts, suffers from latency and often leads to timing jitter. Unless the DSP/processor functions are relatively limited in number or computationally light (which is not the typical case), establishing a consistent data exchange mechanism can become the system's bottleneck. Please note that this is a peculiar condition of this application field, where the computational burden of the DSP/GPP subsystem is, generally, quite heavy and articulated in several concurrent functions.

A more efficient approach is represented by the use of a RTOS, where multi-tasking and efficient data exchange mechanisms, e.g., based on Direct Memory Access (DMA), are built in the OS I/O functions and almost transparent to the user. The DMA channels typically offer multi Mbit/s bandwidths at minimum or even zero latency, which makes their usage compatible with the applications we are discussing.

In systems on chip, like the one described in [3], the data exchange between GPP and FPGA reaches the maximum performance, because the FPGA, I/O resources, and processor natively share the same data bus and memory, operating at the system's clock frequency. Still, efficient function scheduling can pose serious problems to the developer.

Development Tools

Integrated Design Environments (IDE) for multi-platform systems, when available, are often complicated to use, involving simultaneously both the typical tools for firmware/software development and those for HDL driven logic chip design. They require a skilled user that dominates both sides of hardware/software co-design or, possibly, a team of specialized designers concurrently working on the same project.

Some systems, like [5], offer instead a truly integrated design environment, i.e., a common programming framework that unifies the FPGA (hardware) and the DSP/GPP (software/firmware) domains, greatly reducing the application development time. This is possible because the above mentioned issues, i.e., the integration of I/O functions within the FPGA domain and the implementation of efficient data exchange mechanisms between the two domains, take place "behind the scenes," almost transparently to the designer.

Debugging

Debugging a complex digital controller of the type described by Table 7.1 requires a long time and exposes the power converter hardware to damage hazards. In addition, the dynamic behavior of grid optimization algorithms is rather complicated to predict by pencil and paper calculations, while conventional simulation software requires very long computing times to produce useful results in non trivial cases. Both of these issues can be solved by Hardware-In-the-Loop (HIL) real-time simulation, a debugging method that is becoming increasingly popular, as is well exemplified by the large amount of published papers, like [14, 15, 16], that deal with this topic.

Following this approach, a digital controller can be developed and fully tested, without even connecting it to the power converter, until its operational capability has been properly assessed. The first step in the simulation set-up is the development of a discrete time model of the micro-grid, including connection impedances, filters, loads and power converter(s). The model has to be run by a sufficiently powerful computer, so as to allow "real time" signal generation, i.e., the output of analog signals having a sufficiently large bandwidth to replicate accurately the system dynamics with no time dilation. If needed, power amplification can be applied to the computer generated signals, so as to make the emulation of the physical system even more realistic. This is often done when the experimental tests aim at the validation not just of a control board, but also of a power converter or, at least, of the required signal conditioning and acquisition circuitry. If this is not the case, because the analog hardware is itself modeled and computer simulated, the inverter controller can be directly connected to the computer analog outputs.

Several commercial computing hardware solutions are available that allow this type of simulations, like [17, 18, 19] or [20]. In our example, we have considered a *reconfigurable controller module* of the type [20], whose main features are listed in Table 7.4.

Real-time HIL simulations allow not only to accelerate the simulation time, up to three orders of magnitude with respect to a conventional desktop PC simulation, but also to verify the functionality, safely removing the unavoidable bugs, of the very same inverter controller that will be later connected to the physical converter.

Table 7.4: Simulation platform characteristics (cRIO-9082)

Feature	Parameter	Value
Processor	Model Processor Speed	Intel Core i7-660UE 1.33 to 2.4 GHz
Memory	Nonvolatile System	32 GB (min.) 2 GB (min.)
FPGA	Model # Slices # DSP48s	Spartan-6 LX150 23038 180
Network	Network interface	IEEE 802.3 Ethernet
Communication	Port	RS-232, RS-485/422, USB VGA, CAN, MXI-Express
Peripherals	Channel	4 AO, 16-bit, ±10 V, 100 kHz 8 Digital Input/Output [1]

7.4 CONTROLLER VALIDATION

Once its design is completed, the control system needs to be properly validated. For the reasons explained in the Aside 14, validation had better not to be performed directly on the experimental set-up. Instead, different validation methods can be used, prior to the experiment, that allow to effectively debug both the software and the hardware controller components. Only when the controller appears to be bug-free, it is wise to attempt an experimental verification. Therefore, in all the complex applications, like the one we are focusing on, the *step-by-step* validation procedure that is now going to be presented is recommendable. It comprises the successive application of three different test methodologies, that are described in more detail in the following.

As to the definition of the test scenario, we need to devise a sufficiently complex situation to allow the verification of the different control functions, but, at the same time, a sufficiently simple one to provide unambiguous debugging information. As an example, the experiment represented by a step transition from 0 to 2 kW of injected real power, P, with reactive power, Q, set to zero, followed by a second step change where the injected power is transferred abruptly from the real (d) to the reactive power axis (q) can be considered a good starting point. The DER coupling converter will hardly ever experiment such abrupt transients in real life, but, as usual, demanding test conditions are better suited to highlight the achievable performance levels.

In implementing the tests, both at simulation level and experimentally, the grid parameters listed in Table 7.5 have been considered.

Table 7.5: Grid parameters

Parameter	Symbol	Value	
Branch resistance	R_{LINE}	1	Ω
Branch inductance	L_{LINE}	150	μH
Branch length	l_b	260	m
Nominal grid voltage	V_{PCC}	230	V_{rms}
Nominal grid frequency	f_0	50	Hz

7.4.1 SIMULINK© MODEL

As a first verification, the controller can be simply simulated on a PC. In our case we used one whose benchmarking score for the well known Matlab© computing software is the following: [0.22 0.23 0.20 0.25 0.84 1.07]. Prior to do that, a model of the system configuration shown in Fig. 7.1 has to be developed. At the same time, the different control blocks of Fig. 7.2 have to be modeled with adequate detail, emulating fixed point arithmetic where applicable, control delays (due to ADC and DPWM operation), and finite PWM resolution. Once all that is done, results like those shown in Fig. 7.3(a) can be obtained. As can be seen, the basic functionality of the controller can be clearly assessed.

However, this type of system-level simulation simply allows to verify the stability of regulators, the quality of steady-state operation and to estimate the achievable speed of response during transients. None of the practical issues related to the implementation of the controller can be effectively addressed at this level. Indeed, encountering an unexpected dynamic behavior or even instabilities, once the simulation model is turned into a hardware controller, is pretty much a common experience. Furthermore, even a short simulation like that in Fig. 7.3(a), took some minutes to complete on our PC.

7.4.2 REAL-TIME, HARDWARE-IN-THE-LOOP SIMULATION

To better validate control and plant models, to improve the quality of the simulation, reducing its execution time, and to safely test the physical control hardware to be later deployed on the DER coupling inverter, we suggest to use HIL real-time simulation as a second verification step.

As previously explained, with this term we indicate the direct connection between a high performance computer, where a model of the controlled plant is simulated, and the inverter control board, where all the algorithms described above run. Thanks to a multichannel digital to analog converter (DAC) module, the computer is capable of generating high speed analog outputs, that are sampled by the controller as if they were coming from the physical system. Therefore, controller operation is strictly real-time, with no time dilation whatsoever.

A graphical view of the HIL simulation organization for the considered application example, is given in Fig. 7.4.

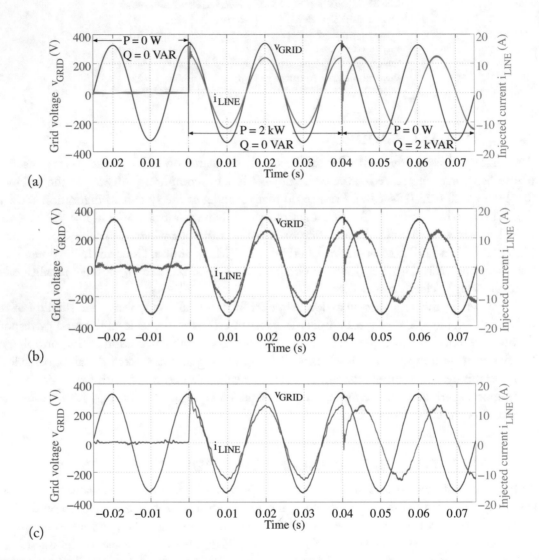

Figure 7.3: Results of control stack test activity: a step reference change is applied successively along *P* and *Q* axis: (a) simulation in Matlab Simulink, (b) hardware in the loop real-time simulation, and (c) experimental test measurements.

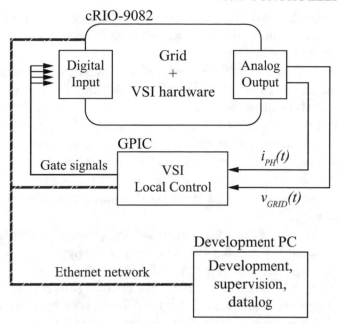

Figure 7.4: Connection between the high-performance computer [20] and the control hardware [5].

There are several practical issues to solve before HIL simulations can be run, as is also pointed out in [15] and discussed in [16]. First, a suitable, i.e., computationally efficient, numerical model of the controlled system has to be set-up.

Sophisticated tools are available on the market that allow to simplify this phase of the model set-up, essentially translating a graphical circuit representation in the corresponding set of equations, that can be then solved by a suitable FPGA synthesized DSP engine. Unfortunately, the cost of these software tools is typically very high.

In a minimum cost approach, instead, the development of the FPGA circuit corresponding to the system's model has to be done manually. The first step is the definition of a discrete time differential equation integrator, formulated to be efficiently implemented by a FPGA circuit. This will be multi-instanced and used to update the state variables of the simulation at run-time. Then, the electrical topology of the system has to be manually specified, applying Kirchhoff voltage and current laws and deriving, by pencil and paper calculations, the set of differential equations that solve the network of interest.

As an example, the most basic integration algorithm for a single reactive circuit element looks like

$$x(k + 1) = x(k) + \alpha u(k), \tag{7.1}$$

where x represents the state variable, α represents the inverse of the value of the electrical parameter (e.g., inductance L) multiplied by the integration step (e.g., $\frac{\Delta T}{L}$), and u is the electrical dual of the considered state variable.

Equation (7.1) shows that FPGA-based, real-time simulation is basically equivalent to numerical fixed step simulation, operated by a particular calculator that, being exactly tailored around the system's differential equations, becomes so fast to be capable of real time signal generation. As such, it maintains all the limitations of fixed step integration of differential equations, in particular a certain lack of robustness and the tendency to generate artifacts due to numerical instability. Unfortunately, implementing more robust or variable step integration methods on a FPGA circuit leads to unacceptable complications in the circuit synthesis, both in terms of needed resources and of computation times, which often leaves the designer with no other option than accepting the above limitations.

Once the passive circuit topology is specified, the power converter has to be modeled. A pure logic-level simulation, idealizing the power switches physical characteristics, is normally sufficient to provide reasonably accurate results.

After compiling the code, the complete plant model runs on the FPGA chip available within the adopted high speed computer [20], so that the set of differential equations that describe the system of Fig. 7.1 is numerically solved by a customized, ad-hoc digital hardware. As implied by (7.1), the integration method is the simple first order Euler. For a limited number of equations, say below 50, the considered hardware capabilities are such that the integration step can be set to $T_{sim} = 100\,\text{ns}$, a relatively small value, that guarantees the accurate replication even of the fast phenomena related to the power converter switching. Fixed point arithmetic is clearly mandatory to preserve computational efficiency, which could pose some numerical stability issues. However, using at least 16 bits for the internal representation of state variables, numerical quantization effects are typically undetectable.

The second issue to be solved is the efficient generation of the analog signals feeding the controller inputs. It is quite common to use relatively slow, e.g., 100 kSample/s, DACs to this purpose. This, however, often poses some limitation on the achievable simulation quality. Indeed, the DAC update delay, together with input and output quantization noises, affect the simulation numerical stability, introducing undesired numerical noise. An example of this unhappy situation is clearly visible on the waveforms presented in Fig. 7.3(b), that have been obtained with a 100 kSample/s DAC. What could be misinterpreted as residual current ripple is indeed a result of some form of numerical simulation instability, as the comparison with the plots in Fig. 7.3(a) suggests. Better results can be achieved adopting the provisions that we are going to describe in the following section.

However, we need to remember that the purpose of real time simulations is to provide a significant test for the controlling hardware. If the simulation artifacts do not impair controller operation, they can be neglected; indeed, they will not be found in the experimental tests (see Fig. 7.3(c)). From this standpoint, even the plots of Fig. 7.3(b) allow us to verify the complete functionality of the controller, practically confirming the results of conventional numerical simulations.

7.4.3   MITIGATION OF REAL-TIME SIMULATION ARTIFACTS

As mentioned in Section 7.4.2, when the DAC update frequency is limited to, say, 100 kHz, simulation artifacts are often generated.

This is due to several reasons. In the first place, a low DAC update frequency results in a measurable update delay, not taken into account in the controller design (which, of course, is correct, since such delay *will not be experienced* by the controller in the experimental tests). An estimation of worst case phase margin reduction determined by such DAC delay is given by the following relation

$$\Delta\phi = -\pi\frac{f_{CR}}{f_{DAC}}\frac{180}{\pi} = 180\frac{f_{CR}}{f_{DAC}}, \tag{7.2}$$

where f_{CR} is the current controller crossover frequency and f_{DAC} is the DAC update frequency.

Equation (7.2) assumes the plant model to be continuous time (it is actually discrete time with 10 MHz sampling frequency) and expresses the phase lag at the desired crossover frequency according to a simple zero order hold (ZOH) approximation of the DAC. In our case, this turns out be equal to 3.6°.

The unwanted phase lag can be better calculated numerically, exploiting (7.3), that represents the plant complete transfer function, i.e.,

$$T_{io}(s) = \frac{V_{DC}\cdot\frac{G_{TI}}{A_r}\cdot e^{-s\,\Delta t_{ctrl}}}{s\,L_{PH} + \frac{1}{s\,C_{PH}}//(s\,L_{GRID} + R_{GRID})}, \tag{7.3}$$

where Δt_{ctrl} is the sum of the AD conversion time (Δt_{AD}), the PI regulator calculation time (Δt_{calc}), given in Table 7.3, and the PWM modulation delay (Δt_{PWM}).

Since the considered implementation adopts a double update PWM, the modulator delay can be estimated to be equal to half the switching period, i.e., to $\frac{T_S}{2}$. Figure 7.5 shows the Bode plots of two discrete time versions of (7.3), both with and without the presence of the DAC delay, that, in the former case, yields a 2.5° additional phase lag.

But the limited DAC update frequency has a much heavier impact on the current controller since, in the worst case, we only generate 2–3 samples in the current ripple run-up (and run-down) phase, which determines a relative error on the average current measurement, $\frac{\Delta I_{phavg}}{I_{ph_{pk}}}$, given by:

$$\frac{1}{3}\frac{\Delta I_{ph_{pk2pk}}}{I_{ph_{pk}}} \leq \frac{\Delta I_{phavg}}{I_{ph_{pk}}} \leq \frac{1}{2}\frac{\Delta I_{ph_{pk2pk}}}{I_{ph_{pk}}}, \tag{7.4}$$

where $\frac{\Delta I_{ph_{pk2pk}}}{I_{ph_{pk}}}$ represents the relative peak to peak inductor current ripple, that, in our inverter, occupies, at its maximum, about 16% of the ADC full scale range. As a result, the worst case

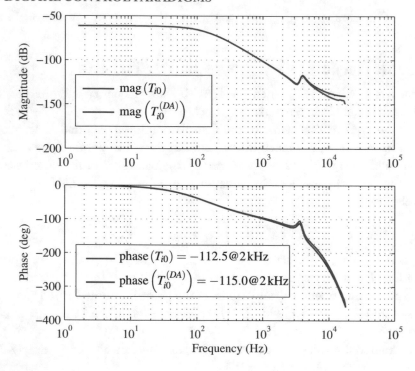

Figure 7.5: Bode plots of the open-loop transfer functions T_{i0} and $T_{i0}^{(DA)}$.

relative error on the average current turns out to be between about 6% and 8%, a non negligible disturbance at the modulation frequency.

Both problems can be solved using faster DACs. For example, it is possible to estimate, by simulations, that the ratio f_{CR}/f_{DAC} should be below 0.05 to give negligible distortion of the generated signals. To achieve that, an upgrade of the DACs to, at least, 2 MSample/s is mandatory.

On the model input side, care must be taken as well. Indeed, the organization of Fig. 7.4 implies a reduction in the actual PWM resolution, as if a lower number of bits were used to represent the duty-cycle. Indeed, although the simulation time step, $T_{sim} = 100$ ns, is adequate to accurately compute the system's state variable dynamics, it is significantly longer than the minimum voltage pulse duration at the PWM adopted resolution. We can express the number of lost bits as

$$n_{b_{lost}} = \log_2 \left(f_{clock} \cdot T_{sim} \right) \tag{7.5}$$

where f_{clock} represents, as usual, the digital PWM counter clock frequency.

The lost bits are partially recovered in the simulation, thanks to the dithering side effect caused by the *asynchronous* operation of the controller and the simulation platform.[1]

Decreasing the simulation time step of the model, reaching the condition $T_{sim} \cdot f_{clk} \leq 1$, it would be possible to read the input gate signals with their full time resolution. However, from the implementation point of view, it is not practical to decrease the simulation time step of the *whole* model to this extent, because this would limit the amount of computable equations and, therefore, the maximum acceptable order of the model.

A more effective method to tackle the problem is by *partitioning* the model, identifying those sections that really need to be simulated at shorter time steps. This approach allows to manage both complexity and time constraints.

Specifically, in the considered case study, the phase inductor L_{PH} is the only component that is exposed to a high speed signal, the full bridge output voltage. Therefore, only its discrete time model actually requires to be computed with a shorter time step.

If the partitioning is properly done, it is possible to achieve better accuracy, while keeping the integration step of the remaining part of the model at lower values. Indeed, the remaining parts of the model experience signals that are dominated by controller dynamics (i.e., bandwidth limited to a few kHz) and, as a result, can be simulated at a slower pace.

To highlight the benefit of partitioning, Fig. 7.6 displays the difference between two successive samples of a low pass filtered version (with 100 Hz cut-off frequency) of the inverter bridge instantaneous output voltage, obtained when a very slow sequential 1 LSB increase of the modulating signal from the valley to the peak of the carrier signal is performed. Being the modulating signal variation well inside the filter bandwidth, ideally, the difference between consecutive filter output readings should approximate the LSB value.

And this is what the figure indeed shows. Thanks to the smaller simulation step of the most critical part of the plant model, it is possible to acquire the average value of the PWM modulated signal almost at its full resolution, equal to to 144 mV.

Figure 7.7 shows the comparison between the partitioned model HIL RT simulation, in Fig. 7.7(a), and the ordinary HIL RT simulation, in Fig. 7.7(b). The improvement in the quality of the results is evident.

7.4.4   EXPERIMENTAL TESTS

The final step of the controller development is represented by experimental tests. In the considered example, we connected the controller to a 3 kVA full bridge VSI (see Table 7.3).

The test results, for the same scenario considered in simulations, are shown in Fig. 7.3(c). Clearly, the power control loops behave exactly as expected. This result demonstrates several achievements of the proposed control development and test methodology.

---

[1]This is a quite tricky effect: because it is not possible to synchronize the controller and the computer where the plant model runs, the DPWM signals are actually averaged by the latter, which tends to mitigate the loss of time resolution.

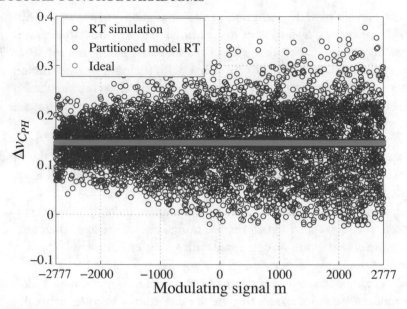

Figure 7.6: Differential increase between two successive samples of a filtered version of the inverter output voltage.

In the first place, we see that local control is performed exactly as in a conventional DSP, or FPGA-based organization. But here, the RT GPP on board the controller and the remote PC supervising the grid are connected by Ethernet and exchange data without interfering with the low level control activities. Once the communication channel bandwidth is defined, thanks to the RTOS, there is no other practical limitation to the information that can be transferred between the two systems in the background, a fundamental requirement to safely operate a large micro-grid experiment (and possibly a real one).

In other words, the multi-platform controller organization is capable of providing both the low level control of a power converter and the required high-level communication and data-logging functions required by the micro-grid control application. The set-up is so flexible that both large and narrow bandwidth communication channels can be emulated, with predefined noise levels, which provides the essential tools for studying the higher level control functions of the stack with very good accuracy.

Besides, the HIL simulation methodology offers a reliable means to verify the functionality of a complex controller, well before serious electrical power comes into play. The simulation results, despite the simple differential equation integration methodology, are good enough to replicate, quite accurately, the physical phenomena that the controller will encounter during the experiments and, as a result, to provide a solid validation of its operation.

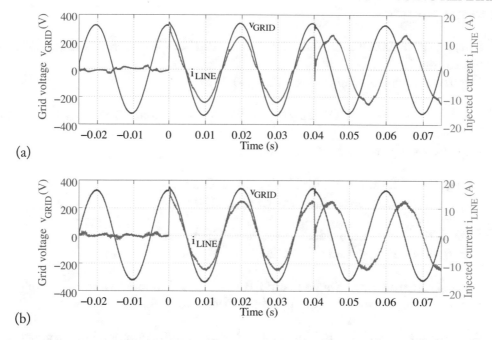

Figure 7.7: Reduction in the simulation artifacts by model partitioning. (a) Partitioned model HIL RT simulation and (b) HIL RT simulation.

7.5   CLOSING REMARKS

The example we have discussed in this chapter is possibly the closest to a real life case of all the examples treated in this book. It well represents what we consider one of the future trends in digital control, the implementation of complex, distributed, multi-platform architectures.

As such, it requires the use of the basic control techniques presented in the early chapters of the book, and even some other ones, that are combined in a multi-functional control stack. We felt it could be useful to visualize how modulation, current and voltage controls, but also the more advanced inverter functions, can be applied to a case of real, practical interest. Its description has allowed us to introduce new control hardware tools and novel design and verification methodologies, such as the hardware in the loop simulations.

We hope this presentation, although quite sketchy and simplified, has given the reader an idea of the complexity involved in the latest development of power converter and digital control applications, stimulating the interest for further reading and research.

REFERENCES

[1] Intersil, "ZL8800 Dual Channel/Dual Phase PMBus ChargeMode Control DC/DC Digital Controller:"
http://www.intersil.com/products/ZL8800 189

[2] Texas Instruments, "UCD9222, Digital PWM System Controller with 4-bit, 6-bit, or 8-bit VID Support:"
http://www.ti.com/product/ucd9222-ep 189

[3] Actel, "Smart fusion customizable System on Chip:"
http://www.actel.com/products/smartfusion/ 190, 198

[4] FOX G20, "Linux Embedded Single Board Computer:"
http://www.acmesystems.it/FOXG20 190

[5] National Instruments, "Single Board RIO - General Purpose Inverter Controller:"
http://www.ni.com/singleboard/gpic/ 190, 194, 196, 198, 203

[6] M. Shahidehpour, S. Pullins and others: "The maturation of Microgrids," *Special issue of the IEEE Electrification Magazine,* Vol. 2, No. 1, March 2014. DOI:10.1109/MELE.2014.2301272 191

[7] H. Kim; T. Yu; S. Choi, "Indirect Current Control Algorithm for Utility Interactive Inverters in Distributed Generation Systems," *IEEE Transactions on Power Electronics,* Vol. 23, No. 3, 2008, pp. 1342–1347. DOI: 10.1109/TPEL.2008.920879. 191

[8] Z. Liu; J. Liu; Y. Zhao, "A Unified Control Strategy for Three-Phase Inverter in Distributed Generation," *IEEE Transactions on Power Electronics,* Vol. 29, No. 3, 2014, pp. 1176–1191. DOI: 10.1109/TPEL.2013.2262078. 191

[9] F. Blaabjerg, K. Ma, Y. Yang, "Power Electronics for Renewable Energy Systems - Status and Trends," 8th International Conference on Integrated Power Systems (CIPS), 25–27 February 2014, ISBN: 978-3-8007-3578-5. 191

[10] M. A. Abusara, J.M. Guerrero, S.M. Sharkh, "Line-interactive UPS for micro-grids," *IEEE Transactions on Industrial Electronics,* Vol. 61, No. 3, March 2014, pp. 1292–1300. DOI: 10.1109/TIE.2013.2262763. 191

[11] C. Buccella, C. Cecati, and H. Latafat, "Digital Control of Power Converters A Survey," *IEEE Transactions on Industrial Informatics,* Vol. 8, No. 3, August 2012, pp. 437–447. DOI: 10.1109/TII.2012.2192280. 194

[12] F. Blaabjerg, R. Teodorescu, M. Liserre, A. V. Timbus, "Overview of control and grid synchronization for distributed power generation systems," *IEEE Transactions on Industrial Electronics,* Vol. 53, No. 5, October 2006, pp. 1398–1409. DOI: 10.1109/TIE.2006.881997. 195

[13] T. Caldognetto, S. Buso, P. Mattavelli, "Digital Controller Development Methodology Based on Real-Time Simulations with LabVIEW FPGA Hardware-Software Toolset", *Electronics Journal,* Vol. 17, No. 2, December 2013, pp. 110–117. DOI: 10.7251/ELS1317110C. 197

[14] C. Wang, W. Li, and J. Belanger, "Real-Time and Faster-Than-Real-Time Simulation of Modular Multilevel Converters using standard multi-core CPU and FPGA Chips," 39th Annual Conference of the IEEE Industrial Electronics Society (IECON), 2013, pp. 5403–5409. DOI: 10.1109/IECON.2013.6700015. 199

[15] C. Graf, J. Maas, T. Schulte, and J. Weise-Emden, "Real-time HIL-simulation of Power Electronics," 34th Annual Conference of the IEEE Industrial Electronics Society (IECON), 2008, pp. 282–2834. DOI: 10.1109/IECON.2008.4758407. 199, 203

[16] A. Hasanzadeh, C. Edrington, N. Stroupe, and T. Bevis, "Real-Time Emulation of a High Speed Micro-Turbine Permanent Magnet Synchronous Generator using Multi-Platform Hardware-in-the-Loop Realization," *IEEE Transactions on Industrial Electronics,* Vol. 61, No. 6, June 2014, pp. 3109–3118. DOI: 10.1109/TIE.2013.2279128. 199, 203

[17] Opal RT Technologies OP5030 - Real-Time Computer: http://www.opal-rt.com 199

[18] RTDS Technologies: https://www.rtds.com 199

[19] Typhoon HIL solutions: http://www.typhoon-hil.com 199

[20] National Instruments, "Compact RIO High performance reconfigurable controller:" http://www.ni.com/compactrio 199, 203, 204

Authors' Biographies

SIMONE BUSO

Simone Buso graduated with a degree in electronic engineering from the University of Padova in 1992. He received his Ph.D. degree in industrial electronics from the same university in 1997. Since 1993, he has been with the power electronics research group of the University of Padova. He is currently a member of the staff of the Department of Information Engineering (DEI) of the University of Padova, where he holds the position of Associate Professor of electronics. His main research interests are in the industrial and power electronics fields and are specifically related to converter topologies, digital control control of power converters, solid-state lighting, and renewable energy sources. Simone Buso is a member of the IEEE.

PAOLO MATTAVELLI

Paolo Mattavelli received his Ph.D. degree in electrical engineering from the University of Padova, Padova, Italy, in 1995. From 1995–2001, he was a Researcher at the University of Padova. From 2001–2005, he was an Associate Professor the University of Udine, where he led the Power Electronics Laboratory. In 2005, he joined the University of Padova, in Vicenza, with the same duties. From 2010–2012, he was a Professor and Member of the Center for Power Electronics Systems (CPES) at Virginia Tech. He is currently a Professor at the University of Padova. His major research interest includes analysis, modeling and analog and digital control of power converters, grid- connected converters for renewable energy systems and micro-grids, and high-temperature and high-power density power electronics. In these research fields, he has been leading several industrial and government projects. Paolo Mattavelli is a Fellow member of the IEEE.